I0057209

Copyright © 2025 by Alec Milner

All rights reserved. No part of this publication may be reproduced, distributed, or transmitted in any form or by any means, including photocopying, recording, or other electronic or mechanical methods, without the prior written permission of the publisher, except in the case of brief quotations embodied in critical reviews and certain other noncommercial uses permitted by copyright law. For permission requests, write to the publisher, addressed "Attention: Permissions Coordinator," at the web address below.

First Article Press

Albuquerque, NM

FirstArticlePress.com

The publisher is not responsible for websites (or their content) that are not owned by the publisher.

The views expressed in this work are those of the author and do not reflect the official policy or position of Sandia National Laboratories, the U.S. Government, or any specific company mentioned herein. All information provided is based on public sources and the author's personal experience.

ISBN 979-8-9942897-0-9 (Hardcover) | ISBN 979-8-9942897-1-6 (Paperback) | ISBN 979-8-9942897-2-3 (eBook)

LCCN: 2025928215

Book design by Alec Milner

Cover design by Alec Milner

Cover Illustration © nnaakk/pixabay

Printed in the United States of America

First Edition: December, 2025

10 9 8 7 6 5 4 3 2 1

THE DEFENSE SECTOR LAUNCHPAD

An Insider's Guide for Aspiring Engineers

Alec Milner

Dedication

To my parents, for nurturing a lifelong passion for engineering and giving me the tools to build a career upon it.

To Lara, for her unwavering patience and support through the long hours of work, school, and writing that made this guide possible.

For my brother, Cole, and for all the young engineers who are determined to build a better and safer future.

And finally, to the warfighters who we, as defense engineers, are always working in the service of.

Foreword

When Alec asked me to write this foreword, I was deeply humbled. Humbled not only by the invitation itself, but by the weight of the topic. To write anything about the defense sector is to attempt to describe something massive. The magnitude of its purpose, the scale of its systems, and the depth of its human contribution defy easy summary. At best, any of us can only offer our limited view—our small reflection on a field whose complexity and importance are nearly boundless.

I have had the privilege of working alongside Alec as he served as a mechanical engineer on an advanced technology project. In that setting, I have seen firsthand the qualities that distinguish great engineers from merely competent ones: character, initiative, and teamwork. Alec embodies those irreducible traits, and this book reflects them throughout. It is not merely a manual on career navigation or a checklist for entering the defense industry. It is a field guide written by someone who understands that engineering is, at its heart, a human endeavor—where success depends as much on integrity, curiosity, and perseverance as on equations or models.

Alec reminds readers that project work is where theory becomes impact. He emphasizes that while academic excellence lays the foundation, it is the discipline and teamwork of applied engineering that build the structures our nation relies upon. His writing conveys what every seasoned engineer eventually learns: that competence is forged through iteration, humility, and the willingness to learn continuously—not through the pursuit of credit, but through contribution to something greater than oneself.

This book neither romanticizes nor trivializes the work of defense engineering. It instead honors the real values that sustain—those who learn by doing, those who collaborate across disciplines, those who build lasting systems and reputations without stepping on others to do so. I asked Alec what prompted him to write this book. He told me it was initiated from material he prepared for his younger brother. For the many of us who do not have an older sibling to put this much applied effort into our career path mentorship, this book offers the next best thing: a thoughtful guide shaped by that very spirit of care, clarity, and experience.

This book will serve young engineers well—not by promising shortcuts, but by revealing the habits and mindset that define enduring success. It is a map drawn from experience and offered with detail. I am proud to introduce it—and proud to have worked beside its author.

— Kendall Key
Chief Engineer, Undisclosed National Security Program, Sandia National Laboratories

CONTENTS

Introduction

The Black Box and the Golden Ticket

If you are an ambitious engineering student nearing graduation, you are standing on a ridge overlooking a vast and sprawling landscape of opportunity. The paths leading down from this ridge are numerous, each promising a distinct challenge, a unique culture, and a different kind of reward. Take a moment to survey the terrain.

One path winds into the vibrant, frenetic world of commercial tech. It leads to the gleaming campuses of Silicon Valley and the disruptive energy of startups, a world that speaks the language of user metrics, agile development, and quarterly earnings. It promises a chance to build consumer-facing products, to "move fast and break things," and perhaps to catch the lightning-in-a-bottle of a successful IPO.

Another path descends into the foundational world of traditional industry and manufacturing. Rather than a digital representation, this is a domain of tangible products that you can feel and touch, from the roar of an automotive assembly line to the quiet precision of a medical device cleanroom. It's a world of process optimization, Six Sigma, and the deep, satisfying hum of a factory floor turning raw materials into the building blocks of our modern society.

A third path leads to the quiet, cerebral world of academia and pure research. This is a journey toward a Ph.D. and a life of intellectual pursuit, dedicated to pushing the very boundaries of human knowledge. It's a life spent in the lab and the lecture hall, contributing to fundamental scientific literature and mentoring the next generation of thinkers.

All these paths are valid, noble, and rewarding, but there is another path, one that can often feel like a black box. It's the world of aerospace and defense; an industry of immense scale and profound importance, yet one that is often shrouded in secrecy, intimidating formality, and a language all its own. It's a world of major prime contractors, prestigious national laboratories, and countless specialized firms in between. You know they build the most advanced machines on the planet, but the way in can seem impossibly complex. It speaks a language all its own, one of acronyms, security clearances, and long-term, mission-critical programs.

I know this terrain because I've been in your exact shoes. I remember the transition from a senior design project sponsored by a Department of Defense (DoD)[1] research lab to my first full-time roles in the industry. I was confident. I had just graduated from a top university with strong internship experiences and a genuine passion for solving complex mechanical problems. I thought my degree had prepared me for everything.

While my academic projects had certainly introduced me to formal processes like Critical Design Reviews (CDRs) and Test Readiness Reviews (TRRs), nothing could have prepared me for their true weight and consequence in a professional environment. Suddenly, these were not academic exercises with a grade on the line, presented to a familiar professor. They were high-stakes events shaping the development of multi-million-dollar, critical national assets, presented to a room of senior engineers and customer representatives, people with decades of experience who would scrutinize every assumption and challenge every data point.

I was immersed in a world where "high-consequence systems" wasn't just a phrase, but a daily reality that guided every engineering decision. The signature on my analysis report or my drawing was not just for a grade; it was a mark of professional accountability for a piece of hardware that had to work, flawlessly, in the most unforgiving environments imaginable. I quickly understood that there was a vast gap between the theoretical knowledge from my coursework and the practical, rigorous, disciplined mindset required to be an effective engineer in this industry. It was a world where failure was not an option, and I realized that my long-term success

---

[1] At the time of writing this book, this was still the correct title. During the process of publishing, the Department of Defense (DoD) has been renamed to the Department of War (DoW) per Executive Order 14347 on September 5th, 2025. Be mindful of this change. This book uses Department of Defense purely for the familiarity of the name to prevent confusion of young engineers who are not already immersed in this world and are not cognizant of the change.

would depend not just on my technical skills, but on my ability to master its unwritten rules, its unique language, and its culture of discipline.

I spent my first year stumbling through this new world, learning its language through trial and error and wishing someone had given me a guide. I made mistakes, missed opportunities, and eventually found my footing, but the process was harder than it needed to be. This book is the map I wish I had to bridge that gap.

It is your comprehensive, step-by-step guide to demystifying the defense sector and landing a rewarding, stable, and purpose-driven career. This isn't a book of abstract theory written by a career coach. It is a practical, actionable playbook written *by an engineer, for engineers*, based on direct experience at the heart of the industry. It is designed to give you an unfair advantage, to take you from the classroom to your first day on the job, and beyond.

Think of this book as your "golden ticket" to that black box. It will provide you with the insider's knowledge to bypass the common mistakes and to present yourself as the exact candidate this industry is desperate to hire: a technically competent, professionally aware, and trustworthy individual.

Our journey together is your mission briefing, broken into seven parts:

Part 1: The Lay of the Land will be your foundational knowledge. You will learn the language of the industry, understand the different types of companies and labs you can work for, and confront the common myths and stereotypes about this unique career path, giving you a powerful case for its mission-driven purpose.

Part 2: The Arenas of Innovation will take you on a deep dive into the five major domains—Air, Land, Sea, Space, and Cyber—to help you find the specific arena that sparks your passion and align your career goals with the concrete, exciting work being done on the frontiers of technology.

Part 3: Charting the Future: Technology, Strategy, and Your Place in It will be your strategic intelligence briefing. We will look over the horizon, exploring the disruptive technological megatrends that are defining the future of defense. We will then translate that foresight into a concrete action plan, detailing the modern, interdisciplinary skillsets you must cultivate to build a resilient, future-proof career.

Part 4: Forging Your Credentials will focus on your college years, showing you how to strategically choose the courses, build the hands-on projects, and land the critical internships that will transform you from a promising student into a must-hire candidate.

Part 5: The Application Arsenal will forge your weapons for the job hunt. I will give you the exact templates and insider strategies to build a keyword-optimized, two-page resume and a compelling cover letter that get you noticed.

Part 6: Navigating the Great Filters will prepare you for the final hurdles. We will walk through the two biggest challenges every candidate faces: the technical interview and the security clearance investigation. We will provide you the frameworks, knowledge, and confidence to pass both.

Part 7: You're In. Now What? will look beyond the offer letter. We will cover the "unwritten rules" of a defense career, guiding you through your first 90 days, setting you up for long-term career progression, and ensuring your success for the first five years and beyond.

Make no mistake: the path ahead is challenging. It demands a higher standard of diligence, precision, and character than many other fields. But the rewards are commensurate with the challenge. It is a career that offers the rare opportunity to work on systems that define the very edge of what's possible, to build a stable and prosperous life upon a bedrock of long-term projects, and to contribute directly to a mission of global significance.

This book is that map. It is designed to be your first piece of mission-critical hardware; a guide to replace uncertainty with a concrete plan, to turn your ambition into a strategy, and to give you the confidence to step into this demanding arena and succeed.

Part 1

The Lay of the Land

Chapter 1

The Case for a Mission-Driven Career

If you're a sharp engineering student nearing graduation, the world is telling you to look west, not just geographically to Silicon Valley, but philosophically toward the fast-paced, consumer-driven world of commercial technology. This path can seem dazzlingly varied. It might be the mythology of a startup, with its promise of hoodies, hockey-stick growth charts, and the chance to disrupt an entire industry from a garage. Or it might be the immense scale and resources of an established FAANG company, a world of complex cloud infrastructure, global user bases, and life-changing stock options. It is a career landscape that casts a long shadow, promising a dynamic world of agile sprints, direct-to-market products, and constant innovation.

Meanwhile, the defense sector can feel like a different universe entirely. The company names of global titans of engineering like Northrop Grumman, Lockheed Martin, Raytheon, and BAE Systems sound formal, even imposing.

This book is your invitation to look behind that curtain. But before we do, we must address the elephant in the room: the perception outside the industry that a career in defense is about "selling your morals for a large salary." This stereotype is born from a misunderstanding of the work itself. While the compensation is competitive, the reality of the mission is far more nuanced. A vast portion of the work is about information, observation, and protection. You may spend your career designing a reconnaissance satellite that provides life-saving disaster relief data, a defensive missile system that protects soldiers and sailors from attack, or a secure communication network that prevents conflicts from starting in the first place. For many, it's a form of patriotic service to build the tools that protect a nation and its allies.

This guide is designed to give you the insider's playbook for this world. We're going to replace the myths with reality and show you why choosing to work on projects of national importance isn't just a good career move; for the right kind of person, it's the best one.

Myth vs. Reality: The Defense Sector

Let's start by tackling the biggest misconceptions head-on.

The Myth	The Reality
"The technology is old and outdated."	The defense industry operates on the bleeding edge, often a decade ahead of the commercial world. It is the unsung engine of global innovation. The Internet, GPS, and jet engines all have their roots in defense research.
"The pay is bad compared to Big Tech."	While a tech startup's offer might have a higher starting salary on paper (often with volatile stock options), the defense sector offers a powerful total compensation package. This includes highly competitive salaries, robust 401(k) matching, excellent health insurance, and often a pension, which are a rarity in the modern economy that provides incredible long-term financial security.
"The work is slow and boring."	This is a common but incomplete picture. The *pace* on large, multi-decade programs is necessarily deliberate due to the immense complexity and high stakes. However, a significant and exciting part of the industry operates at a sprint. When an urgent, mission-critical need arises (a "strategic offset"), teams are formed to develop and deploy a solution as quickly as humanly possible. This work is incredibly fast-paced, agile, and focused.
"There's no room for	This is a nuanced issue. While large, established programs require a rigorous and disciplined process, the industry

creativity; it's also has a massive and growing need for rapid innovation.
all rigid bu- Internal R&D groups, Skunk Works®-style teams, and the
reaucracy." entire defense startup ecosystem thrive on a "move fast
and find what works" mindset. The creativity comes from
solving impossible problems, whether that's through a
flash of brilliance in a rapid prototype or a deep, meticulous insight on a formal program.

"It's impossi- It's a different process, but they are in constant need of
ble to get a job new talent. They simply look for different signals than a
there." tech startup does. This book will teach you those signals.

This is an outdated stereotype, but it requires a careful
understanding. The industry is rapidly evolving, with
many companies adopting more flexible dress codes and
"It's a stuffy, work styles, especially in R&D and software environ-
old-fashioned, ments. However, this flexibility should not be mistaken
'gray suit' cul- for a lack of professionalism. The focus is shifting to value
ture." outcomes over appearances, but a high standard of professional conduct, communication, and respect is always expected. It's a culture where your work speaks for itself, but
it must do so from a foundation of professional integrity.

"You're just a
tiny cog in a Programs are massive, but you will own critical compo-
giant ma- nents and systems. Your contribution is tangible and has a
chine." direct impact on the final product's success.

A Career with Gravity: The Power of Purpose

The single greatest differentiator between a commercial and a defense career is the *why*. In a commercial company, the ultimate goal, no
matter how it's phrased, is often tied to quarterly earnings or shareholder
value. The problems are often fascinating, but the mission is business.

In the defense sector, the mission is unambiguous: to design, build,
and sustain the systems that provide security and stability for a nation and
its allies. It is a world where the end-user is not a consumer, but a soldier, a

sailor, an airman, or a guardian, and the "user experience" is measured in mission success and survivability. This reality imbues the daily engineering work with a tangible and profound sense of gravity. The technical challenges you solve have a direct line to a larger, national purpose.

Consider the specifics:

The materials engineer isn't just selecting a polymer for a consumer product's casing; they are qualifying a novel composite for the leading edge of a hypersonic vehicle, knowing it must withstand the searing, multi-thousand-degree temperatures of atmospheric reentry without failing.

The electrical engineer isn't just designing a circuit board for a laptop; they are creating the receiver for an electronic warfare system, knowing it must be hardened to function flawlessly through a barrage of hostile jamming and an electromagnetic pulse (EMP) that would instantly destroy conventional electronics.

The mechanical engineer isn't just designing a housing to be waterproof; they are analyzing a submarine's hull to withstand the crushing, bone-breaking pressure of the deep ocean, where a single microscopic flaw in a weld could lead to catastrophic failure.

The software engineer isn't just debugging an app; they are writing the guidance, navigation, and control (GNC) code for a missile defense interceptor, where a timing error of a few milliseconds could be the difference between a successful intercept and a national catastrophe.

This is not to say one is better than the other, but they are fundamentally different. Work in the defense industry has a weight that provides a powerful, intrinsic motivation that is hard to replicate elsewhere. You are not just solving a technical problem; you are delivering a capability that must work, without fail, under the most extreme conditions imaginable.

The Weight of Consequence

We must also address the reality of what "mission success" often implies. In the defense industry, we use clinical terms like "kinetic effects," "target neutralization," and "lethality." But as an engineer, you must understand that you are designing systems that may, one day, be used to destroy a target or take a life in defense of the nation. This is not a video game. The hardware you build has real-world consequences.

This gravity is what separates this profession from designing a toaster or a social media app. It requires a somber and serious ethical framework. You must be comfortable with the mission. For some, working on defensive systems like missile interceptors is a perfect fit, while offensive weaponry feels too heavy. For others, providing the warfighter with the most effective, overwhelming firepower is a moral imperative to end conflicts quickly. There is no "right" answer, but you must have your own answer. This industry demands engineers who approach their work not with casual detachment, but with a profound sense of moral responsibility and a commitment to ensuring these systems work exactly as intended, precisely when needed, and never by accident.

The Sandbox of Advanced Technology

One of the most persistent, and most incorrect, stereotypes of the defense industry is that its technology is outdated. The truth is more complex and far more exciting. While the technology on any given program may be "behind" where the engineers wish it could be against the ever-advancing threat, the industry as a whole is in a constant state of reaching for what's next. It is an ecosystem perpetually living in the future, solving problems that the commercial market won't encounter for another decade.

The list of technologies that made the leap from a classified program to your daily life is staggering. The Internet began as ARPANET, a DoD project to create a resilient network. GPS was a military satellite constellation before it guided you on your daily commute. Even the microchip itself was propelled into mass production by the guidance computer of the Apollo program and the Minuteman missile.

Today, that legacy of pioneering innovation continues. As an engineer in this industry, you will be on the front lines of fields that are still in the realm of theory for the commercial world:

Artificial Intelligence & True Autonomy: This is not just a self-driving car navigating a well-mapped highway. This is the deep, complex AI and machine learning required to enable a team of unmanned systems to navigate a hostile, unmapped, and communications-denied environment, making decisions on its own to complete a mission.

Hypersonics & Directed Energy: While commercial aviation chases incremental fuel efficiency, you will be designing vehicles that withstand the physics-bending reality of flight at over five times the speed of sound, or working on the high-energy laser and high-power microwave systems that represent a fundamental shift in defensive technology.

Space Systems & Satellite Technology: You will be engineering the next generation of resilient GPS, reconnaissance, and global communication satellites, designing the systems that serve as the backbone of modern navigation and intelligence from the ultimate high ground.

Advanced Materials & Manufacturing: This is the science of creating the impossible. You will be engineering the exotic ceramic matrix composite, stealth coatings, and additively manufactured (3D printed) metal alloys that allow systems to perform under unimaginable temperatures and stresses.

Energetics & Advanced Propulsion: This is the chemical engineer's realm. It involves the research and development of the next generation of high-performance, stable solid rocket propellants, advanced explosive materials, and the chemical processes required to produce them safely and reliably.

Next-Generation Semiconductors: The demand for extreme performance in tiny spaces for radar and electronic warfare is driving the development of exotic semiconductor materials like Gallium Nitride (GaN) and Silicon Carbide (SiC), long before they become mainstream.

Quantum Technology: While still in its early stages, the race to develop quantum sensors for navigation (that don't rely on GPS) and quantum computing for code-breaking is being heavily funded and explored within national labs and defense R&D centers.

You won't just be reading about these breakthroughs in a tech journal; you will be in the lab, at the computer, and on the manufacturing floor bringing them to life.

Bedrock Stability in a World of Quicksand

The commercial tech world, for all its excitement, runs on a volatile cycle of boom and bust. The defense sector, in contrast, offers a unique spectrum of stability.

The foundation of the industry is built upon massive, multi-billion-dollar Programs of Record. The contracts to build the next fleet of submarines or fighter jets that are backed by the full funding of the U.S. government and can span decades. For engineers working on these programs, the stability is almost unimaginable in the private sector. It is a world of long-term planning, deep expertise, and unparalleled job security, like building a cathedral that will stand for a century.

However, the industry is not exclusively made up of these marathons. There is also a vibrant and growing ecosystem of exploratory, high-risk, high-reward contracts. These are the faster-paced projects funded by organizations like the Defense Advanced Research Projects Agency (DARPA) and other R&D groups. This work, often pursued by internal rapid-prototyping teams, smaller specialty firms, and defense startups, is focused on developing the next breakthrough technology. While these contracts are shorter and carry more risk, they offer an exciting, dynamic environment that rivals the pace of any tech startup.

This dual nature allows you to choose your path. You can build a career on the bedrock of a major, long-term program or thrive in the fast-paced world of cutting-edge R&D. Either way, the industry as a whole provides a foundation that allows you to excel.

You will be able to build deep, lasting expertise. In a commercial environment where employees often switch companies every 18-24 months, it can be difficult to develop true mastery of a complex system. The defense industry encourages the opposite. The long-term nature of the work allows, and in fact requires, you to become a true Subject Matter Expert (SME). You will have the time to go deep, to understand a system from its initial concept through its design, testing, and decades of sustainment. You will become the person who knows not just *how* it works, but *why* it was designed that way, a level of knowledge that builds immense professional value and deep personal satisfaction.

You will have the confidence to plan your life. The stability of the government-funded ecosystem provides a powerful sense of security that is increasingly rare. You are not constantly worried about the next round of venture capital funding drying up or a sudden shift in market sentiment leading to layoffs. This stability is profoundly liberating. It is the confidence to buy a home and put down roots in a community. It is the freedom to start a family, knowing your excellent health insurance and steady paycheck will be there for them. It is the ability to plan for a comfortable retirement, contributing to a robust 401(k) and often a pension, without the constant background anxiety of industry volatility.

You will have the freedom to choose your pace and find your balance. The famous 9/80 schedule (every other Friday off) and respect for the 40-hour work week are common, providing a sustainable work-life balance that is a rarity in many industries. This is the marathon. However, if you are an engineer who thrives on intensity, you can absolutely find a sprint. There are numerous programs and R&D groups focused on developing quick-turn solutions for critical, time-sensitive technological gaps. This work is an "all-aspects sprint" that is demanding, fast-paced, and incredibly rewarding. The unique strength of a defense career is the ability to often choose the pace that fits your personality and career stage, all while being supported by generous benefits and, often, a pension.

This book is your training plan for navigating this rich and varied career. Now that we have made the case for the "why," the following chapters will systematically teach you the "how."

Chapter 2

The Defense Industry Ecosystem

To an outsider, the defense industry can seem like a monolithic entity, a handful of giant corporations with impenetrable walls. In reality, it's a complex and fascinating ecosystem; a deeply interconnected industrial web of thousands of players, each with a distinct role, culture, and rhythm. Understanding this web is the first step in targeting your job search effectively, and more importantly, in finding the specific environment where your talents and personality will truly thrive.

At the center of this web is the customer: the U.S. Department of Defense (DoD). They define the mission, set the high-level requirements for a new capability (e.g., "we need a new long-range bomber with stealth capabilities"), and provide the funding through the congressional budget.

To understand how the industry responds to these needs, let's use an analogy that every engineer can appreciate: building a complex, one-of-a-kind custom house.

The Prime Contractors

The Prime Contractors are the giants of the defense world, the titans whose names are synonymous with the industry. In our analogy, the Prime is the General Contractor and the Lead Architect for the entire project.

When the customer (the DoD) issues a set of requirements for a new "house," they award the master contract, often worth tens of billions of dollars and spanning decades, to a single Prime. The Prime is now responsible for everything: translating the customer's high-level needs into a detailed architectural blueprint (systems engineering), managing the immense budget and schedule, and ultimately, handing over the keys to a finished product that works perfectly. They build the core structure, the foundation and frame of the house (the airframe, the ship's hull), but for the thousands of specialty systems inside, they rely on a vast network of partners.

Let's open the front doors on these titans. These aren't just companies; they are industrial empires, each with their own culture, history, and crown jewels. These are the system integrators who manage the largest and most complex programs imaginable.

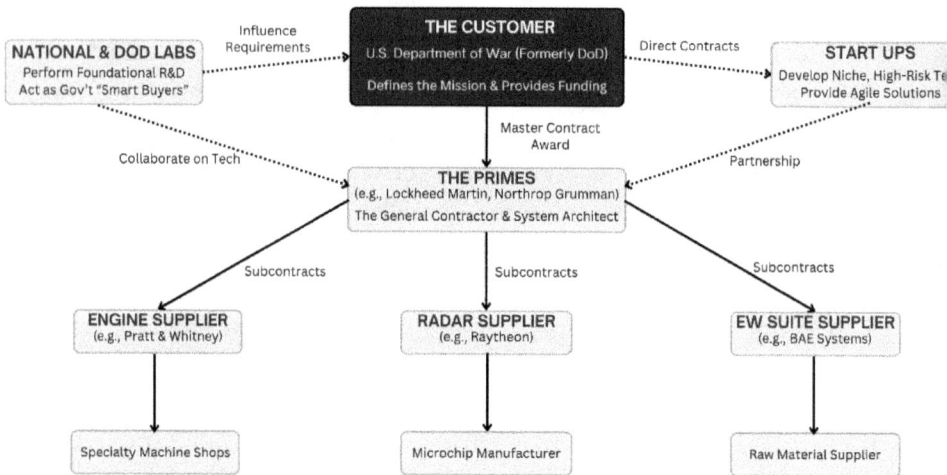

Figure 1 The Defense Industrial Ecosystem; This hierarchy shows a highly generalized view representing the flow of requirements and funding of a program.

1. Lockheed Martin

Perhaps the largest and most diversified defense contractor, Lockheed Martin, is a titan in nearly every domain, a true multi-disciplinary giant.

Aeronautics: This is their most famous division, home to the legendary Skunk Works® advanced development group in Palmdale,

California. Skunk Works is famous for developing the nation's most secret and advanced aircraft, like the U-2 Dragon Lady, SR-71 Blackbird and the F-117 Nighthawk stealth fighter. Today, this division produces iconic aircraft like the F-22 Raptor and the F-35 Lightning II, as well as the workhorse C-130 Hercules transport.

Rotary and Mission Systems (RMS): Through their ownership of Sikorsky, they are a world leader in helicopters, producing the iconic Black Hawk and Seahawk helicopters that are the backbone of Army and Navy aviation. RMS also has a massive naval systems portfolio, including the Aegis Combat System, the integrated network of radars and missiles that serves as the "shield" of the U.S. Navy's destroyer fleet.

Missiles and Fire Control (MFC): Headquartered in Orlando, this division is a leader in precision munitions, tactical missiles, and the advanced sensor and targeting systems that go with them. They are also a major player in the development of hypersonic systems.

Space: A massive player in satellite technology for decades, Lockheed Martin Space builds everything from the nation's resilient GPS constellation to deep-space exploration probes for NASA, like the Orion spacecraft.

2. Raytheon Technologies (RTX)

Following its landmark merger with United Technologies, Raytheon Technologies (RTX) became a true powerhouse, a giant with deep expertise spanning the full spectrum of sensors, munitions, and both commercial and defense aerospace systems.

Raytheon: This is the core of their defense business. They are the undisputed global leader in missile systems, producing a vast arsenal including the Tomahawk cruise missile, the AMRAAM air-to-air missile, and critical defensive systems like the Patriot missile defense system. They are also a world leader in radars, from the

massive AN/SPY-6 that equips Navy warships to the advanced
AESA radars on fighter jets.

Collins Aerospace: A major provider of both commercial and military avionics. If you've flown on a modern airliner, you've almost certainly interacted with their products. They build cockpit displays, communication systems, landing gear, and countless other critical aircraft components.

Pratt & Whitney: One of the "big three" jet engine manufacturers in the world. They produce the engines for a huge portion of the world's commercial airliners (like the Airbus A320neo) and are a primary engine provider for military fighters, including the F-35's powerful and incredibly complex F135 engine.

3. Northrop Grumman

A leader in stealth, autonomous systems, and space technology, Northrop Grumman is the name synonymous with many of the nation's most advanced and secretive programs.

Aeronautics Systems: This division is synonymous with stealth. They produced the iconic B-2 Spirit stealth bomber and are now leading the development of its revolutionary successor, the B-21 Raider.

Mission Systems: This division is a leader in the "brains" behind the platform, building advanced radars, sensors, and electronic warfare suites.

Space Systems: A major force in space, responsible for building assets like the James Webb Space Telescope for NASA and leading the nation's strategic land-based missile programs, ensuring the readiness of the ICBM fleet.

Defense Systems: This includes a diverse portfolio, from battle management systems that network the entire force to cutting-edge undersea capabilities. A key example is their Manta Ray prototype, an extra-large unmanned underwater vehicle (UUV) designed for long-range, long-duration autonomous missions that is expanding the frontiers of naval warfare.

4. Boeing Defense, Space & Security

The defense arm of the world's largest aerospace company, Boeing has a legendary portfolio of military aircraft and is a key player in space and autonomous systems.

Military Aircraft: Boeing is responsible for a huge portion of the U.S. military's current air fleet, including the legendary F-15 Eagle and F/A-18 Super Hornet fighters, the P-8 Poseidon maritime patrol aircraft (a modified 737), and the KC-46 aerial refueling tanker.

Space and Launch: A major manufacturer of military and commercial satellites for decades.

Phantom Works: Boeing's advanced prototyping division, their equivalent to Lockheed's Skunk Works, which develops next-generation concepts for air, sea, and space.

5. General Dynamics

General Dynamics is a powerhouse on the ground and at sea, producing some of the most critical and iconic platforms for the U.S. Army and Navy.

Land Systems: This division is the primary builder of the U.S. Army's armored fleet, including the M1 Abrams main battle tank, often called the most capable tank in the world, and the versatile Stryker family of wheeled combat vehicles.

Electric Boat & Bath Iron Works: Through these two legendary shipyards, General Dynamics is a cornerstone of the U.S. Navy. Electric Boat, in Groton, Connecticut, is one of only two shipyards in the nation capable of building nuclear-powered submarines, including the Virginia-class attack submarines and the new Columbia-class ballistic missile submarines, the nation's most survivable strategic deterrent. Bath Iron Works in Maine is a primary builder of the Arleigh Burke-class guided-missile destroyers.

L3 Harris Technologies

Created through a mega-merger of L3 Technologies and Harris Corporation, and recently expanded by acquiring Aerojet Rocketdyne, this company is a dominant force in the unseen backbone of the military: communications, electronics, and propulsion.

Frequently referred to as the "Sixth Prime," especially following its acquisition of Aerojet Rocketdyne, L3Harris has cemented itself as the nation's premier provider of propulsion and rocket engines for everything from space launch vehicles to hypersonic missiles. Beyond propulsion, they are a dominant force in space and airborne systems, manufacturing the sophisticated sensors, avionics, and electronic warfare suites that empower platforms like the F-35. On the ground, they dominate the market for critical soldier systems; if a soldier is speaking on a secure radio or operating through a night vision device in the dark of the battlefield, there is a very high probability that equipment was built by L3Harris.

Leidos

Leidos represents the "services and solutions" side of the industry. While they do build hardware (including pioneering work on autonomous Navy vessels like the Sea Hunter), their massive footprint is in Intelligence, IT modernization, and scientific support.

Leidos serves as the giant behind digital modernization, architecting and maintaining the massive networks that connect the Pentagon and global military bases. In the realms of intelligence and cyber, they employ thousands of analysts and engineers who work directly alongside government customers to process critical data and defend these networks. Furthermore, they have established themselves as a major player in the health and civil sectors, managing the complex healthcare and logistics systems that support the human element of the force.

6. The "Near-Primes" & Technology Integrators

There is a unique tier of companies that sits between the platform-building giants and the traditional subcontractors. These are massive, multi-billion-dollar organizations that act as Primes on complex subsystems (like propulsion, communications, or IT infrastructure) and are increasingly competing directly with the "Big 5" for major contracts.

What it's like to work at a Prime: You work on history-making programs with immense resources. The benefits are top-tier, and the career paths are well-defined. Because of the scale, your role might be highly specialized, focusing on a specific component of a massive system. The pace is deliberate and structured, governed by the formal processes required to manage immense complexity.

The Subcontractors

No General Contractor can be the world's best electrician, plumber, and roofer simultaneously. They hire specialists. Similarly, no Prime Contractor can build every single component of a modern weapon system. They hire a vast network of subcontractors.

The Tiered System:

Tier 1 Subcontractors: These are the master specialists hired directly by the Prime. For our fighter jet, Raytheon might be the "master electrician" responsible for designing and building the entire, incredibly complex radar system. BAE Systems might be the "security specialist" for the electronic warfare suite, and Pratt & Whitney is the "HVAC expert" for the jet engine. These are large, highly capable companies in their own right, and they often act as Primes on other programs.

Tier 2 & 3 Subcontractors: These are the suppliers who provide parts *to the specialists.* A smaller company that makes high-performance, radiation-hardened microchips sells them to Raytheon (the electrician), who then integrates them into the final radar system. A specialty machine shop might build a specific valve that goes into Pratt & Whitney's engine.

What it's like to work there: At a subcontractor, you become a deep subject matter expert in your niche. Your work is often more focused on a single technology. You may have a broader range of responsibilities within your specialty (designing, testing, and building the entire "electrical system"). The companies are often smaller and can be more agile than the Primes, with less bureaucracy.

This tiered system exists to leverage specialized expertise and manage immense risk. It ensures that every single component of a multi-billion-dollar program is designed by a team that is the best in the world at what they do.

The R&D Labs

The prime contractor is building the house with today's best practices, but where do the next-generation technologies—like smart-grid electrical systems, advanced composite building materials, or futuristic security sensors—come from? They are born in the dedicated research and development labs.

This category of the ecosystem is where the foundational science and deep research that enable future capabilities take place. For engineers who are passionate about R&D and want to be close to the core mission, the government's own network of laboratories offers a unique and highly rewarding career path. These institutions are broadly divided into two families: those that work directly for the military branches and those that work on the nation's most sensitive strategic challenges under the Department of Energy.

1. The Department of Defense (DoD) Service Laboratories

Think of these labs as the dedicated R&D and engineering brain trust for each branch of the military. As a civilian engineer working here, your "customer" is the soldier, sailor, airman, or guardian. Your job is to understand their needs, develop the technology to meet those needs, and act as the government's "smart buyer" when working with the big Prime Contractors.

> **Naval Sea Systems Command (NAVSEA) and its Warfare Centers (e.g., NUWC, NSWC):** Located in places like Newport, RI, and Carderock, MD, these centers are the heart of ship and submarine technology. Engineers here work on everything from advanced hull designs and hydrodynamics to the complex combat systems, sonar, and torpedoes of the submarine fleet.

> **Naval Air Systems Command (NAVAIR):** Focused on naval aviation. At sites like Patuxent River, MD, engineers work on the

unique challenges of launching jets from aircraft carriers, naval aircraft propulsion, and the sophisticated electronics that go into maritime patrol aircraft.

Naval Research Laboratory (NRL): The Navy's corporate lab, with a broader, more scientific focus. This is where foundational technologies like GPS were born.

Air Force Research Laboratory (AFRL): This is the massive R&D organization for both the Air Force and Space Force, with directorates across the country. Engineers at Wright-Patterson Air Force Base, OH, might work on next-generation jet engines and aerospace systems. At Kirtland Air Force Base, NM, the focus is on space systems, directed energy (lasers), and satellites.

U.S. Army Combat Capabilities Development Command (DEVCOM): This is the Army's primary technology leader. At Aberdeen Proving Ground, MD, engineers work on everything from advanced materials and ballistics to command and control software. At the Ground Vehicle Systems Center (GVSC) in Detroit, MI, the focus is squarely on the future of Army tanks, robotic combat vehicles, and advanced powertrains.

What it's like: You work directly alongside uniformed military personnel and are deeply connected to the end-user. The work is mission-focused, and as a federal employee, you get incredible job security and benefits.

2. The NNSA National Laboratories

This unique and prestigious part of the ecosystem is centered around the labs of the National Nuclear Security Administration (NNSA), a semi-autonomous agency within the Department of Energy. While the NNSA's core mission is to maintain the safety, security, and reliability of the U.S. nuclear deterrent, this single, profound responsibility has created a scientific and engineering enterprise of unparalleled capability.

To solve the nation's most difficult nuclear challenges, these labs have become world leaders in a vast array of fields, from high-performance supercomputing to materials science and advanced manufacturing. As a result, a large and growing portion of their work is dedicated to supporting the Department of Defense on critical non-nuclear national security projects, applying their unique expertise to a wide range of conventional defense challenges.

It's important to note that the labs listed below are just a few prominent examples from the full, nationwide network of National Laboratories, each with its own unique specializations.

These labs include but are not limited to:

Los Alamos National Laboratory (LANL) - Los Alamos, NM: The birthplace of the atomic bomb, LANL continues its primary mission of designing the nuclear "physics packages" for the nation's deterrent.

Lawrence Livermore National Laboratory (LLNL) - Livermore, CA: The second of the nation's nuclear design labs, LLNL shares the primary mission with Los Alamos. It is also home to the National Ignition Facility (NIF), the world's largest laser, used for fusion research.

The National Engineering and Manufacturing Labs:

Sandia National Laboratories - Albuquerque, NM & Livermore, CA: While the design labs like Los Alamos handle the nuclear physics, Sandia's primary mission is to engineer everything else: all the non-nuclear components and systems that make the deterrent a safe, secure, and reliable whole. This makes them world experts in high-consequence engineering, ensuring absolute safety and reliability. This expertise is now applied to a huge portfolio of non-nuclear work for the DoD, including the design of advanced satellite payloads, hypersonic flight systems, and advanced radar development.

Oak Ridge National Laboratory (ORNL) - Oak Ridge, TN: While not an NNSA lab, this major DOE Office of Science lab is a powerhouse in materials science, neutron science, and high-performance computing. Its deep expertise in manufacturing is crucial to the defense sector. A prime example of cross-lab synergy is the CAMINO (Center for Advanced Manufacturing and Innovation) initiative, a collaboration between Sandia and Oak Ridge to accelerate the development and adoption of next-generation manufacturing technologies for national security applications.

What it's like: A hybrid culture that blends the intellectual freedom of a top-tier university with the profound responsibility of a critical national security mission. An engineer here might spend one day working on a deeply classified strategic program and the next day collaborating with a university on foundational materials research. The labs tackle the "grand challenges" that are too big or too high-risk for any single company, making them a fantastic option for engineers who love deep, foundational R&D that spans a wide range of national security problems.

The Startups

While the Primes and National Labs represent the established bedrock of the industry, a new and disruptive force has emerged in recent years: the venture-backed defense startup (e.g., Anduril Industries, Shield AI, and Palantir). In our analogy, these are the boutique architects and high-tech gadget inventors. They don't try to build the whole house. They focus on doing one thing, often software, AI, or autonomous systems, ten times better and faster than anyone else. They aim to disrupt the traditional building process with radical new ideas.

What it's like: The fastest paced and most dynamic environment. You work on a small, agile team with immense autonomy. It's a higher-risk, higher-reward path focused on rapid prototyping and bringing a Silicon Valley mindset to the defense world.

Why This Industrial Web Matters to You

This ecosystem is your map of opportunity. Understanding its structure transforms your job search from a random walk into a targeted campaign. It allows you to understand that a career at a "Tier 2 Subcontractor" might offer deeper technical focus than at a Prime. It reveals that a role at a "DoD Lab" provides a closer connection to the military end-user. It shows that a career at a "Startup" offers a different balance of risk and reward than one at a "National Lab."

Use this chapter not as a list to be memorized, but as a strategic guide. Use it to find the specific culture, pace, and technical challenges where your talents and personality will truly thrive. It is the intelligence you need to find your perfect place within this vast and fascinating industrial web.

Chapter 3

A Brief History of the Defense Industry

To truly understand the culture and structure of the modern American defense industry with its emphasis on formal processes, long-term perspective, and unique, symbiotic relationship with the government, you have to understand where it came from. The ecosystem we described in the last chapter was not inevitable. It was forged in the heart of global conflict, shaped by the existential pressures of the Cold War, and deliberately architected by decades of policy. Knowing this history is not an academic exercise; it is your first piece of critical intelligence. It is the key that unlocks the 'why' behind the culture you are about to enter from its formal processes to its corporate structure. Understanding this DNA will give you the context to navigate your new environment not as a confused outsider, but as an informed professional.

The World Before

Prior to the 20th century, the concept of a permanent, professional "defense industry" did not exist in the United States. The nation, protected by two vast oceans, held a deep-seated skepticism of a large, standing military. Armaments were produced in government-owned "arsenals," like the Springfield Armory in Massachusetts or the Watervliet Arsenal in New York. These were facilities focused on craftsmanship and slow, deliberate production of rifles and cannons. In times of war, like the Civil War, the government would issue temporary contracts to private factories to produce weapons, but these relationships would dissolve as soon as the conflict ended. There was no permanent partnership, no industry whose primary business was preparing for the next war. That all changed with the cataclysm of the Second World War.

World War II and the Birth of the Arsenal of Democracy

When war broke out in Europe in 1939, the United States was dangerously unprepared. The U.S. Army was the 17th largest in the world, smaller than that of Portugal, and its soldiers trained with wooden mockups of machine guns. The attack on Pearl Harbor on December 7, 1941, was a terrifying wake-up call that the oceans would no longer be a sufficient shield.

President Franklin D. Roosevelt's response was to call upon the one force he knew could turn the tide: the immense, untapped power of American private industry. In a famous fireside chat, he declared that the U.S. must become the "great arsenal of democracy," providing the tools for the Allies to win the war. This set in motion a national mobilization on a scale the world had never seen, coordinated by the new War Production Board. The government effectively told the titans of private industry, companies like Ford, General Motors, Chrysler, and Boeing, to stop making consumer goods and start making war materiel.

The results were staggering and arguably won the war. The statistics are almost incomprehensible today. Ford Motor Company's Willow Run plant, a sprawling complex built on peaceful farmland in Michigan, went from breaking ground to producing a four-engine B-24 Liberator bomber nearly every 60 minutes at its peak, rolling them off a mile-long assembly line. Chrysler, a maker of passenger cars, became a primary builder of Sherman tanks, applying their mass-production genius to heavy armor. Shipyards, led by the brilliant industrialist Henry J. Kaiser, began producing Liberty ships, the workhorse cargo vessels of the war, using revolutionary pre-fabrication and welding techniques. At their peak, they were completing a ship from keel to launch in just a few weeks; one famous example was built in an astonishing four days and 15 hours.

This massive, successful partnership, where the government provided the funding and the requirements and private industry provided the innovation and manufacturing might, became the foundational blueprint for the modern defense sector. It proved that a public-private collaboration could achieve monumental engineering and production feats. But even then, it was seen as a temporary alliance, expected to dissolve once peace returned.

The Cold War and the First Offset

The world did not return to peace. The "Iron Curtain" descended across Europe, and the Cold War, defined by the decades-long technological and ideological, created a new, terrifying reality. The advent of the atomic bomb and long-range bombers meant that a future war could be lost in a matter of hours, not years. The core strategic problem was that the Soviet Union and its satellite states had a massive conventional military advantage in Europe. The U.S. and its NATO allies could not hope to match them tank for tank, or soldier for soldier.

To counter this overwhelming numerical inferiority, President Eisenhower's administration developed what would become known as the First Offset Strategy. The plan was simple and terrifying in its logic: to "offset" the Soviet conventional advantage with the overwhelming technological superiority of nuclear weapons. This strategic necessity drove the engineering of the "Nuclear Triad," a massive, multi-decade effort to build and perfect the three pillars of a survivable deterrent force, ensuring that no surprise attack could disarm the United States. This effort created the bedrock of the modern aerospace industry:

Long-Range Strategic Bombers: Led by the Air Force, this required building a fleet of aircraft that could fly across continents. This drove the development of giants like Boeing's B-52 Stratofortress, an engineering marvel that is still in service today, more than 70 years after its first flight.

Land-Based Intercontinental Ballistic Missiles (ICBMs): Perhaps the defining technical challenge of the era. The Minuteman program required breakthroughs in solid-fuel rocketry, guidance systems that could hit a target on the other side of the world, and control systems that had to work with absolute certainty. Companies like TRW and Boeing became leaders in this field.

Nuclear-Powered Ballistic Missile Submarines (SSBNs): Arguably the most complex machines ever built. Led by Admiral Hyman G. Rickover, the "Father of the Nuclear Navy," companies like General Dynamics Electric Boat developed submarines that could stay hidden in the depths of the ocean for months at a time, providing the ultimate survivable deterrent.

This era codified the methodical, process-driven approach to engineering. When the consequence of failure in a system like an ICBM or a nuclear submarine was not just losing a battle, but potentially triggering a global catastrophe, a new level of rigor, documentation, and what we now call systems engineering was born. Here lie the origins of the formal, process-heavy culture that still characterizes large programs today.

The Second Offset and the Dawn of Smart Weapons

By the 1970s, the strategic landscape had shifted again. The Soviet Union had achieved rough nuclear parity, erasing the decisive advantage of the First Offset. The U.S. was once again faced with the challenge of a massive Soviet conventional force in Europe. This led to the Second Offset Strategy, a brilliant pivot championed by visionary figures like Secretary of Defense Harold Brown. The goal was to again use technology to offset the Soviet numbers, but this time with a new generation of smart, conventional weapons guided by the power of the microchip.

This strategy was the primary driver of technological innovation for the next two decades, and it created the high-tech defense industry we recognize today. It was a fusion of advanced sensors, networked computing, and precision guidance. This is the era that gave birth to:

Stealth Technology: In the deep secrecy of Lockheed's legendary Skunk Works®, engineers like Ben Rich developed the F-117 Nighthawk, the first operational stealth aircraft, designed to be functionally invisible to Soviet air defense radars. It was a revolutionary leap in aeronautical science.

Precision-Guided Munitions (PGMs): The creation of laser-guided bombs and GPS-guided missiles (like the Tomahawk cruise missile) fundamentally changed warfare. A single weapon could now destroy a target that would have previously required hundreds of bombs, reducing risk and increasing effectiveness exponentially.

Advanced ISR (Intelligence, Surveillance, and Reconnaissance):
The development of sophisticated spy satellites and incredible platforms like the U-2 and the Mach 3 SR-71 Blackbird gave the U.S. the ability to see and hear deep into enemy territory with unmatched clarity, providing a decisive strategic advantage.

The Creation of DARPA: To ensure the nation was never again surprised like it was by Sputnik, the Defense Advanced Research Projects Agency (DARPA) was created. Its mission was to pursue high-risk, high-reward research that would lead to revolutionary capabilities. It was DARPA that funded the creation of ARPANET, the precursor to the Internet, as well as early research in GPS, stealth, and countless other technologies.

It was the overwhelming success of these Second Offset technologies in the first Gulf War in 1991 where the world watched live as precision weapons performed with incredible accuracy that truly demonstrated the power of this new era of warfare.

"The Last Supper" and the Great Consolidation

Just as the Second Offset was proving its dominance, its primary reason for being vanished. The Soviet Union collapsed in 1991, and the Cold War was over. The defense industry, which had been structured for a global superpower competition, faced an existential crisis. The "peace dividend" led to massive, immediate cuts in the defense budget. The industry was suddenly oversized for a world with only one superpower, and many firms faced bankruptcy.

In 1993, then-Secretary of Defense Les Aspin invited the CEOs of the dozens of major defense and aerospace companies to a dinner in the Pentagon that would become legendary. This event is known in industry circles as "The Last Supper." His message was blunt and historic: the Pentagon could no longer afford to support all of them in a competitive landscape. He projected future procurement budgets on a screen, showing a steep decline, and famously told the assembled executives to "consolidate or be liquidated." It was a direct instruction to merge and shrink.

This set off a frantic wave of mergers and acquisitions throughout the 1990s as companies scrambled to survive. It was a high-stakes game of musical chairs that fundamentally reshaped the industry. Giants swallowed giants.

Martin Marietta merged with Lockheed to form Lockheed Martin. Lockheed Martin then acquired General Dynamics' famous Fort Worth aerospace division, the maker of the F-16. Boeing acquired McDonnell Douglas, the legendary builder of the F-15 Eagle and F/A-18 Hornet. Northrop acquired Grumman (maker of the F-14 Tomcat), and later the advanced technology company TRW. Raytheon bought Hughes Aircraft, a leader in electronics and missiles.

This period of intense consolidation directly explains why, today, we have a small number of giant, multi-billion-dollar Prime Contractors instead of the dozens of smaller, fiercely competitive firms that existed during the Cold War.

The Modern Era, The Third Offset, and the Digital Age

The post-9/11 era brought another major shift. The focus turned from state-on-state conflict to intelligence, surveillance, reconnaissance (ISR), and counterterrorism. This fueled the growth of a new kind of defense contractor, one focused not just on big hardware platforms, but on software, data analysis, network security, and unmanned systems like the Predator drone.

More recently, the re-emergence of strategic competition with near-peer adversaries has presented a new challenge. The stealth and precision technologies of the Second Offset are no longer a unique American advantage. This has led the Pentagon to champion a Third Offset Strategy. The goal is, once again, to use technology to gain a decisive advantage, but this time the focus is on leveraging autonomy, artificial intelligence, and human-machine teaming. Understanding this strategy is critical, as it is the primary force shaping the technological megatrends, as well as the engineering job opportunities of the next several decades, as we will explore in detail in Chapter 12.

This new strategy drives the primary innovation in the industry today, and it explains the immense focus on:

Artificial Intelligence & Machine Learning: Developing algorithms that can analyze vast amounts of data from sensors to identify threats and recommend actions faster than any human.

Autonomous Systems: Building the unmanned air, sea, and ground vehicles that can operate in collaborative swarms.

Networked "Kill Webs": Creating the resilient communication and data networks (often called Joint All-Domain Command and Control, or JADC2) that can connect any sensor to any shooter on the battlefield in real-time.

This is the world you are about to enter: an ecosystem of giant, consolidated Primes born from the Cold War; a deep network of specialized subcontractors; and a new, dynamic layer of startups, all racing to invent the future of warfare. This industry has a long memory. Its culture is a direct reflection of this dramatic history. When you encounter a process that feels rigorous and slow, you are seeing the ghost of the Nuclear Triad, where the consequence of a mistake was unimaginable. When you see the immense focus on AI and data, you are seeing the Third Offset Strategy in action. Understanding this story isn't just about knowing the past; it's about decoding the present and having the tools to build the future.

Chapter 4

Cracking the Code

Every profession has its own language, a specialized vocabulary that allows its members to communicate with precision and efficiency. Doctors speak of "stat orders" and "differential diagnoses." Lawyers speak of "torts" and "injunctions." This specialized language is a tool, a form of intellectual shorthand that conveys complex ideas quickly and unambiguously. In the world of aerospace and defense, this language is more than just jargon; it is a critical system, engineered with the same rigor as the hardware itself, designed for success and survival in a high-consequence environment.

This language is built on a foundation of acronyms, classifications, and programmatic terms, and to an outsider, it can be bewildering, a seemingly impenetrable wall of alphabet soup. It is one of the first and most significant cultural barriers a new engineer will face. Sitting in your first design review, hearing a senior leader ask, "What's the TRL of the COTS components in this IPT's baseline, and does it align with the SOW for the EMD phase?" can be an intimidating and humbling experience.

But this language was not created to be confusing. It was created to be precise. In an industry where a misunderstanding between a software engineer in California and a hardware engineer in Maryland could lead to a catastrophic failure on a billion-dollar satellite, ambiguity is the enemy. This language is a form of engineering itself; a system that is designed to ensure that every single person on a program of thousands is speaking about the same requirement, the same process, and the same standard of quality with absolute clarity.

Mastering this language isn't just about memorizing a list of acronyms; it's about understanding the culture, the priorities, and the structure of the industry. When you can use these terms correctly and confidently in a career fair conversation, an interview, or a team meeting, you are sending a powerful signal. You are showing that you have done your homework, that you are serious about this career, and that you already understand the world they live in. It is a key that unlocks a deeper level of professional conversation and demonstrates that you are ready to be a contributing member of the team. This chapter is your essential language and culture guide, your Rosetta Stone for the world of defense engineering.

The sheer volume of terms and acronyms in this chapter can seem overwhelming, like trying to drink from a firehose. The most important piece of advice is this: do not try to memorize them all at once. This chapter is not a test; it is a reference. It is your foundational dictionary, the guidebook you can return to again and again throughout your early career.

Your goal is not rote memorization, but gradual immersion. In your first few months on the job, keep this chapter handy. When you hear a term you don't understand in a meeting or read it in a formal document, look it up. The act of looking it up in the context of a real-world problem will cement its meaning in your mind far more effectively than any flashcard. Ask your mentor or a trusted teammate, "I heard the term 'TRR' in the meeting today. Can you explain what that means in the context of our project?" This shows curiosity and a drive to learn.

The Language of the Program

This is the language of how projects are born, funded, and managed. It is the vocabulary of the "big picture," the framework that governs the entire engineering enterprise. Understanding these terms is key to understanding the context of your daily work. They are the words that program managers and chief engineers live by, and your ability to understand them will show a level of professional maturity that is rare in a new hire. It demonstrates that you see beyond your specific task and understand how your work contributes to the larger mission of the program.

Acquisition Lifecycle

Definition: The formal, multi-stage process by which the DoD develops and fields a new system, moving from its initial conception to its eventual retirement decades later.

Why It Matters: This is the master roadmap for every single defense program. It's a slow, deliberate, and highly structured process designed to manage immense technical risk and the expenditure of billions of taxpayer dollars. It ensures that a program proves its technology is mature before major funding is committed and that the design is solid before manufacturing begins.

How You'll Encounter It: You will be hired to work on a program in a specific phase. Early on, in Technology Maturation & Risk Reduction (TMRR), the work is highly experimental and R&D focused. The bulk of engineers work in the Engineering and Manufacturing Development (EMD) phase, which is where the detailed design, analysis, and prototyping happens. Understanding which phase your program is in is critical to understanding its current priorities and the type of work you will be doing.

The Defense Acquisition Lifecycle

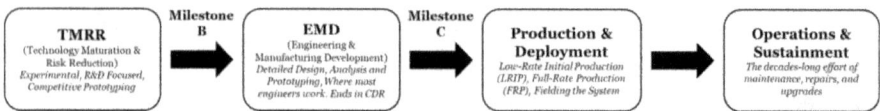

Figure 2: A Simplified View of the Defense Acquisition Lifecycle. This formal, multi-stage process, defined in DoD Instruction 5000.02, guides a program from its early technological development through final production and fielding. For engineers, the journey typically begins in the Technology Maturation & Risk Reduction (TMRR) phase. A successful program then passes Milestone B, the critical decision point to enter the Engineering & Manufacturing Development (EMD) phase, where the bulk of detailed design work occurs. Following Milestone C, the program proceeds into Production & Deployment and its decades-long life in Operations & Sustainment.

RFP (Request for Proposal)

This is the starting gun for a new program. The government issues an RFP detailing the high-level requirements and objectives for a new system they need.

The RFP is the genesis of all work. It triggers a massive, high-stakes competition between the Prime Contractors. Companies will spend millions of dollars and assign their best engineers for months, or even years, to write a detailed, often thousand-page proposal that explains their technical solution, their management plan, and their cost estimate. Winning a major RFP can secure a company's future for a decade or more.

Early in your career, you are unlikely to work on a proposal, but you will hear senior leaders talk about "the RFP" for the next big contract the company is trying to win. It is the lifeblood of the business.

Contract

A contract is the legally binding agreement between the government and the winning company that formally kicks off a program.

The contract type dictates many of the program's incentives and constraints. A Fixed-Price contract means the company has agreed to deliver the system for a set price, which incentivizes intense cost control and efficiency. A Cost-Plus contract means the government pays the company's costs plus an agreed-upon fee, which is often used for high-risk R&D where the final cost is unknown and the primary goal is technological breakthrough.

While you won't be negotiating the contract, its terms will influence the decisions your lead engineer makes every single day. The pressure to control costs or the freedom to innovate is often a direct result of the contract type.

Program of Record

A Program of Record is an acquisition program that has been formally approved and funded by Congress in the national budget.

This term signifies stability and a long-term commitment from the U.S. government. A "Program of Record" is a real, funded program with a multi-year plan, not an experimental project that could be canceled tomorrow. It is a major pillar of the nation's defense strategy.

In an interview, being able to say, "I'm interested in working on a major Program of Record like the B-21," shows you understand the difference between a fleeting research project and a major, long-term engineering effort. It signals that you are thinking about a stable, long-term career.

Design Reviews (PDR, CDR)

Design Reviews are the formal, milestone-based reviews where the engineering team must present and defend its design to the customer (the government) and senior leadership. (These are often referred to collectively as 'major design reviews.' While PDR and CDR are the most common terms, you may also encounter others like FDR, or Final Design Review, which often serves a similar purpose to a CDR before production.)

These are the high-stakes "final exams" for a design; they are the gates that control the flow of a multi-billion dollar program. Passing a PDR (Preliminary Design Review) proves that your high-level concept is feasible and that you have a credible plan, unlocking the funding for detailed design. Passing a CDR (Critical Design Review) is the ultimate test, proving that your detailed design is complete, backed by exhaustive analysis, and ready for the immense expense of manufacturing. A failed CDR can put a program months or even years behind schedule and can be a career-defining event for its leaders.

As a junior engineer, your life will revolve around preparing for these reviews. You will be tasked with creating the specific slides that display the detailed analysis results, the CAD views, and the margin summaries that your lead engineer will present. You will participate in countless internal "murder board" practice sessions where senior engineers grill your team to find any weaknesses. During the review itself, you will sit in the audience, heart pounding, waiting for your slide to come up, ready to answer a direct, probing question from a customer's chief engineer about the specifics of your analysis.

IPT (Integrated Product Team)

These are small, multi-disciplinary team of engineers (mechanical, electrical, software, etc.) responsible for a specific part of a larger program (e.g., the "Wing IPT" or the "Radar IPT").

The IPT is the fundamental organizational unit of a large program. It is your team, your "home base." It is the mechanism that breaks down an impossibly complex system into manageable chunks.

You will be assigned to an IPT, and the other members of that team will be your primary collaborators. You will attend your IPT's weekly meetings, and the IPT Lead will be your day-to-day technical supervisor.

SOW (Statement of Work)

A Statement of Work is the section of the contract that details the specific work activities, tasks, and deliverables the contractor is required to perform. Consider this the master "to-do" list for the program. This contractual document defines what "done" looks like.

When your lead engineer assigns you a task, that task is ultimately derived from a requirement in the SOW. Understanding the SOW for your part of the program gives you a clearer picture of the overall goals.

WBS (Work Breakdown Structure)

A Work Breakdown Structure is a hierarchical decomposition of all the work to be executed by the program team, broken down into smaller and smaller packages.

The WBS serves as the primary tool for organizing the work and, most importantly, for tracking cost and schedule. It provides a structured way to manage the budget for a massive program.

Every task you work on, and every single hour you charge to your timesheet, will be assigned to a specific WBS number. It is a fundamental part of the daily and weekly business rhythm of the company.

Prime Contractor & Subcontractor

The Prime is the company that wins the main contract from the government and is responsible for integrating the entire system.

The Subcontractors are other companies hired by the Prime to provide specific components or expertise.

This relationship defines the industrial web. No single company can build everything. The Prime relies on its trusted subcontractors to provide critical, specialized systems like engines, radars, or landing gear.

You will frequently be in meetings with engineers from subcontractor companies, working to solve the integration problems between their component and your system. You will learn to work collaboratively across corporate boundaries.

Sustainment

Sustainment is defined as the decades-long effort of maintaining, repairing, and upgrading systems that are already in the field, long after the main production line has shut down.

It is a massive and highly stable source of engineering work, often with a larger budget than the initial development program. A fighter jet designed in the 1970s is still flying today because of the continuous work of sustainment engineers.

You may be hired as a sustainment engineer, a challenging and highly respected role where you will be tasked with solving complex obsolescence issues (like finding a replacement for a 30-year-old microchip) and integrating modern technology onto decades-old platforms.

The Language of the Engineer

These are the terms that describe the engineering culture and the day-to-day work. They are the vocabulary of the engineering process, the words that describe *how* a complex system is designed and validated. Using these correctly on your resume and in interviews is critical to proving your technical credibility and demonstrating that you understand how professional, high-consequence engineering is done.

Systems Engineering

Systems Engineering is a holistic, interdisciplinary approach to designing and managing complex systems, typically from the top down. It is the discipline that focuses not just on the individual components, but on the interfaces and interactions between them to ensure the entire system accomplishes its mission.

In a system as complex as a fighter jet, you cannot have the engine team working in a vacuum, separate from the fuselage team or the flight control team. It is the dominant and most important engineering philosophy in the entire industry. Systems engineering is the discipline that ensures all the individual pieces of the puzzle fit together, that the requirements are properly managed, and that the final, integrated system meets the customer's needs.

You will be a member of an IPT, working alongside engineers from other disciplines. You will be expected to think like a systems engineer, to constantly ask, "How does my component affect the others?" You will participate in the formal systems engineering process, which includes the other terms in this section like requirements management, V&V, and trade studies.

Requirements & Requirements Traceability

Requirements are the specific, verifiable capabilities a system must have. They are almost always written as "shall statements" (e.g., "The aircraft shall fly at Mach 2.0"). Traceability is the meticulous, and often arduous, process of linking every single design feature, piece of code, and test case back to a specific, parent requirement.

This is the bedrock of Verification & Validation (V&V). It is the formal, contractual proof that you built what you were contracted to build. If a feature exists that doesn't trace back to a requirement, it is "gold plating" and a waste of money. If a requirement exists that isn't verified by a test or analysis, you have not fulfilled your contract. Traceability is the unbreakable chain of logic that connects the customer's initial desire to the final, proven product.

Your first task as an intern or new hire might be to help a senior engineer perform a traceability analysis in a requirements management tool like Jama Connect or IBM DOORS. This might involve reading a 500-page system specification and ensuring that every single "shall statement" is correctly linked to a corresponding design document and a verification test.

V&V (Verification and Validation)

V&V is the formal process of testing and analysis to prove a system works. Verification asks the question, "Did we build the system right?" (Does it meet every single detailed specification in the requirements document?). Validation asks the more fundamental question, "Did we build the right system?" (Does it actually solve the customer's high-level problem and accomplish the mission?).

Here, you prove your design works. V&V acts as the formal process of moving from theory and simulation to proven, physical reality. The V&V plan is a massive undertaking that is planned years in advance.

As a junior engineer, you will be deeply involved in V&V. You will write and execute formal test procedures for your component. You will analyze the data from those tests. And the results of your work will be used as the official, contractual evidence that the requirement has been successfully verified.

Configuration Management (CM)

This is a rigorous, formal process for controlling and documenting every single change to a design. The goal is to ensure that everyone on the program is always working from the same, approved "baseline" design.

In a program with thousands of engineers working for a decade, CM is the discipline that prevents absolute chaos. It ensures that the manufacturing team is not building a part based on an outdated drawing that the design team changed last week. It provides a complete, auditable history of every decision and every change made to the design.

You will not be able to make a change to a formal design, no matter how small, without submitting an Engineering Change Request (ECR). This process will feel slow and bureaucratic at first, but you will come to appreciate it as the heartbeat of professional, high-consequence engineering.

Trade Study

A Trade Study is a formal, documented process for comparing different design options against a set of weighted criteria (like performance, cost, risk, and "the -ilities"). This is how major engineering decisions are made and justified to the customer. It provides a data-driven, logical, and unbiased basis for your design choices, moving beyond "gut feelings" or personal preference.

As a junior engineer, you may be tasked with gathering the data for a trade study. For example, your lead might ask you to research three different types of sensors for a new application and to create a comparison matrix of their performance, cost, maturity (TRL), and SWaP.

SME (Subject Matter Expert)

Pronounced "smee," an SME is a person with deep, universally recognized expertise in a specific technical area (e.g., a "heat transfer SME" or a "guidance algorithms SME"). SMEs serve as the technical backbone of the company, the keepers of the corporate memory, and the ultimate mentors. They are the ones who solve the problems that no one else can.

You will frequently seek out SMEs for advice on your most difficult problems. Being able to say in a meeting, "I've already consulted with the thermal SME on this," gives your work instant credibility and shows that you are leveraging the company's most valuable resources.

"The -ilities"

This is common engineering slang for the critical non-functional requirements that determine a system's real-world usefulness: *Reliability, Maintainability, Manufacturability, Sustainability, Affordability.* A design that only focuses on raw performance is a failed design if it's impossible to build, too expensive to operate, or can't be repaired in the field. "The -ilities" are what separate a clever prototype from a successful product.

In every single design review, you will be expected to speak to how your design addresses "the -ilities." How easy is it to manufacture your part? How will a soldier in the field replace a broken component? These questions are just as important as "Does it meet the performance spec?"

Technology Readiness Level (TRL)

Technical Readiness Level is a scale from 1 to 9, originally developed by NASA, used to measure the maturity of a technology. It provides a common, universal language for assessing and communicating technical risk. A program that relies on a TRL-2 (a concept on a whiteboard) is much riskier than one that relies on a TRL-8 (a system that has been proven in a real environment).

You will hear program managers and chief engineers talk constantly about the TRL of different components. A major goal of an R&D program is often to "mature a technology from a TRL-3 to a TRL-6," which is the level generally required to transition it into a formal Program of Record.

Interface Control Document (ICD)

An Interface Control Document defines the precise mechanical, electrical, and logical interface between two components or systems, often managed by two different teams or even two different companies. The ICD is the "contract" between two engineering teams. It is the single source of truth for how two systems will connect and interact. A well-written ICD prevents the catastrophic "integration phase" problems where two components physically don't fit or electrically don't talk to each other.

You will spend a great deal of time in meetings called Interface Control Working Groups (ICWGs), meticulously reviewing and updating ICDs line by line to ensure that your component will successfully "talk to" the component it is plugging into.

Peer Review

Peer Reviews are the process of having your design, your analysis, or your code reviewed by your fellow engineers to find errors before it is formally released. It is the cornerstone of technical quality, a form of collaborative error-checking. No engineer is perfect, and a good peer review process is a safety net that catches mistakes. It is a sign of a healthy engineering culture.

This will be a constant part of your work life. You will both give and receive peer reviews weekly. It requires a thick skin and a commitment to improving the team's product, not just defending your own work. It is one of the primary ways you will learn from the more experienced engineers on your team.

The Language of the Shop Floor

This is the practical language that connects the world of digital design to the world of physical hardware. This is where the rubber meets the road, where your pristine CAD model is transformed into a real, tangible object made of metal, composites, and wires. Understanding these terms is not just an academic exercise; it will earn you immense respect from the technicians, machinists, and manufacturing engineers who have the difficult job of actually building what you design. Speaking this language proves that you are not just a "paper engineer," but someone who understands and respects the entire engineering lifecycle.

Red Line

Red Lines are physical, red-penned markups of an engineering drawing, typically done by a senior engineer, a customer representative, or a manufacturing technician, indicating a required or suggested change.

The "red line" is a powerful and traditional form of direct, informal feedback. In the world of formal Configuration Management, a red line is not an official change, but it is the critical first step. The Red Line serves as the tangible record of a problem found or an improvement suggested during a review or on the shop floor.

As a junior engineer, one of your very first tasks will almost certainly be to take a "red-lined" drawing from your lead engineer and incorporate the changes into the official CAD model. You will then use those changes to create a new revision of the drawing and submit it through the formal Engineering Change Request process.

Traveler

This is a packet of paperwork that physically travels with a piece of hardware through the entire manufacturing and inspection process. The traveler is the birth certificate and the complete, auditable history of a physical part. It contains the drawing, the step-by-step manufacturing instructions, and, most importantly, the sign-offs and stamps from every single person who worked on or inspected the part at each step. In a high-consequence industry, this traceability is paramount.

When a part fails during testing, the very first thing a Failure Review Board will do is request the traveler for that specific part. They will use it to trace the part's history, to see if a step was missed, or to identify which machine or operator was involved in its creation.

Material Review Board (MRB)

A Material Review Board is a formal meeting of engineering, quality, and manufacturing representatives to decide the fate of a part that has been made incorrectly and does not conform to the drawing. This is where the hard, and often expensive, decisions are made. Perfection in manufacturing is impossible, and non-conforming parts are a reality. The MRB is the formal process for dealing with this reality in a disciplined way.

As you become more responsible for a design, you will be called into MRB meetings to provide the engineering disposition. You will be asked, "The hole on this part was drilled 0.005 inches out of position. From an engineering perspective, can we still use it?" You will then have to perform the analysis to prove whether the non-conformance is acceptable ("Use As Is"), can be repaired ("Rework"), or if the part must be thrown away ("Scrap").

As-Built vs. As-Designed

As-Designed is the perfect, idealized state of your CAD model and its drawing. As-Built is the reality of the physical part, with all its real-world manufacturing variations, which are documented in inspection reports that are part of the traveler. The difference between these two states is the source of many engineering problems. The discipline of tolerance analysis is dedicated to ensuring that a system will still function even when all the parts are at the extremes of their "as-built" condition.

You will spend a great deal of your time as a junior engineer comparing "as-built" inspection data against the "as-designed" model to ensure that the parts you are receiving from the shop floor are acceptable.

COTS (Commercial Off-the-Shelf)

A component that is bought from a commercial catalog (like a bolt from McMaster-Carr or a microchip from Digi-Key) rather than being custom-designed and manufactured for the program is considered to be a COTS part. Using COTS components is a critical strategy for saving time and money. However, it also introduces risk, as you have no control over the component's design or manufacturing process, and it may not be designed to survive a harsh military environment.

You will be involved in countless "make vs. buy" trade studies. You will spend a great deal of time researching COTS components to see if they can meet the program's requirements, and if you choose to use one, you will often be responsible for designing and executing the tests to "qualify" it for use.

GD&T (Geometric Dimensioning and Tolerancing)

GD&T is the symbolic language used on engineering drawings to define the precise form, orientation, and location of features on a part, beyond just its basic dimensions. It is an unambiguous way to communicate incredibly complex tolerance requirements between the engineer and the machinist. A well-executed GD&T scheme ensures that parts will fit together and function correctly, every single time.

You must become fluent in reading and applying GD&T. Your drawings will be reviewed by senior engineers and machinists, and your ability to apply GD&T correctly is a direct measure of your competence as a mechanical designer.

DFM (Design for Manufacturability)

Design for Manufacturability is the engineering process of actively designing parts that are easy, cheap, and reliable to manufacture. A brilliant, high-performance design that is impossible or incredibly expensive to make is a failed design. DFM is the art of designing with the manufacturing process in mind.

You will participate in DFM reviews with manufacturing engineers. They will look at your design and give you feedback like, "This internal corner is too sharp for our standard tools; can you add a radius?" or "If you change this tolerance, we can make the part in half the time."

Tooling

Tooling is the custom jigs, fixtures, molds, and other equipment that are created for the specific purpose of manufacturing and assembling your part. This is often incredibly expensive and can take a long time to design and build. A design change that requires new tooling is a major cost and schedule impact to a program.

You will work with tooling engineers to ensure that your design is compatible with their manufacturing plan. You will hear program managers agonize over the cost and lead time of "long-lead tooling."

First Article Inspection (FAI)

A First Article Inspection is a detailed, formal, and exhaustive inspection of the very first part produced by a new manufacturing process. The FAI is a high-value verification that the manufacturing process is capable of producing a part that meets the design intent. Every single dimension, tolerance, and feature on the drawing is measured and verified.

As the design engineer, you will be a key participant in the FAI process. You will review the inspection data and be responsible for formally signing off that the first article is acceptable before full-rate production can begin.

Non-Conformance

This is a formal term for a part or feature that is out of tolerance and does not meet the drawing's specifications. This is the trigger for the MRB process that was previously covered.

A quality inspector will find a non-conformance on your part and will write a formal "non-conformance report." This report will be the primary input to the MRB meeting where you will have to help decide the fate of your part.

The Language of Risk and Performance

This is the language of engineering management and program leadership. These are not just technical terms; they are the vocabulary of decision-making. Understanding these concepts shows that you are thinking not just about the design of your individual component, but about the health, performance, and success of the program as a whole. Fluency in this language is a clear sign to senior leaders that you are a future leader yourself.

Technical Performance Measure (TPM)

TPMS are the key technical attributes of a system that are formally tracked throughout the design process to ensure the program is on a path to success. TPMs are the vital signs of the program. For an aircraft, the most important TPMs are almost always Weight, Range, and Payload. For a radar system, they might be Detection Range and Power Consumption. These are the handful of critical metrics that, if they go in the wrong direction, can put the entire program at risk.

In every single major design review, you will see a series of charts showing the current, predicted value of the key TPMs versus their required, "must-not-exceed" value. As a junior engineer, your analysis will directly feed into the tracking of these critical metrics. When you perform a weight analysis on your component, that number rolls up into the master TPM for the entire system.

Risk Matrix

A Risk Matrix is used to formally manage program risks, plotting the likelihood of a risk (a potential problem) occurring against the consequence (the severity of the impact) if it does. It is the primary tool that program leaders use to prevent program failure. It allows them to focus their limited time, budget, and engineering resources on the most important problems. A "red" risk (high likelihood, high consequence) will get daily attention from the Chief Engineer. A "green" risk (low likelihood, low consequence) will be watched but not actively worked.

In team meetings, you will hear your lead engineer talk about the team's top risks and the plans to "burn them down." You may be assigned a task that is part of a formal risk mitigation plan, such as building a prototype to prove that a new, risky technology actually works.

Margin

Margin is the engineering "safety buffer" in a design. It is the difference between how strong or capable a part is required to be and how strong or capable it is actually designed to be. It is a measure of the design's robustness and its ability to handle unexpected loads, future growth, or manufacturing imperfections. A design with "zero margin" is a design that is predicted to work perfectly, but only if everything goes exactly as planned, which it never does. A design with "negative margin" is a design that is formally predicted to fail.

The direct output of your stress analysis will be a "margin of safety." You will spend a huge portion of your career in a constant battle to maintain "positive margin" across all of your designs. In design reviews, you will be expected to present your margin calculations with rigorous, well-documented proof.

SWaP-C (Size, Weight, and Power - Cost)

SWaP-C are the four most critical, and often conflicting, design constraints for almost any component, especially in the Air and Space domains. It is the language of trade-offs. You can almost always make a component smaller, lighter, or more powerful, but it will almost always come at the expense of one of the other parameters, or it will increase the cost. Engineering is the art of balancing these competing constraints.

You will hear this acronym in almost every single meeting. Your lead engineer will challenge you: "This design is great, but it's too heavy. How can you reduce the Weight?" or "The electrical team says we don't have enough Power for that new feature. What are our options?"

Trade Space

The Trade Space is the set of all possible design solutions for a given problem, defined by the trade-offs between different performance parameters. It provides a structured way to think about and explore all the potential solutions, not just the first one that comes to mind. A formal exploration of the trade space is the foundation of a good trade study.

In the early stages of a design, you will be asked to "explore the trade space." This might involve creating a series of simple concept models or analyses to understand the relationship between, for example, the thickness of a part and its weight versus its strength.

Root Cause Analysis (RCA)

Root Cause Analysis is the formal, structured problem-solving method used to identify the fundamental, underlying cause of a failure, not just its immediate symptoms. Fixing a symptom is a temporary patch; fixing the root cause is a permanent solution. RCA is a critical discipline for ensuring that a failure never happens again.

When a test fails, a formal Failure Review Board (FRB) will be convened, and they will use a structured RCA process (like a "fishbone diagram" or the "5 Whys") to find the true root cause. As a junior engineer, you will be a key participant in this process for your components.

Failure Mode and Effects Analysis (FMEA)

A FMEA is a systematic, proactive, and often spreadsheet-based analysis to identify all the potential ways a design could fail ("failure mode") and what the consequences of those failures would be ("effects"). It is a tool for proactive risk mitigation. It forces you to think like a pessimist, to anticipate problems before they happen, so you can design your system to be more robust.

You will be required to complete a FMEA for your design as part of the formal systems engineering process. You will sit in a room with senior engineers and brainstorm all the ways your design could break, and then you will have to show how your design mitigates the most critical of those failure modes.

Test Readiness Review (TRR)

A TRR is a formal review, similar to a PDR or CDR, held before a major, expensive, or high-risk test. It ensures that you don't waste millions of dollars or risk destroying a priceless piece of prototype hardware because of a simple mistake. It is a formal "go/no-go" decision point.

Before your component can go into the vibration lab or the thermal vacuum chamber, you will have to present at a TRR. You will need to prove that your test article is ready, that your test procedures are complete and have been peer-reviewed, and that all the necessary resources are in place.

"Long Pole in the Tent" (Critical Path)

This is common industry slang for the single biggest technical or schedule risk that is driving the critical path for the entire program. This is the problem that keeps the Chief Engineer and the Program Manager awake at night. It is the one thing that has the potential to derail the entire project.

You will hear this phrase constantly in leadership meetings. "What's the long pole in the tent for the software team?" The team or the component that is the "long pole" will receive an immense amount of scrutiny and resources.

Baseline

Baselines are the formally approved, configuration-controlled versions of a design, a document, or a plan. The baseline is the single, official point of departure for all future work. It is the "stake in the ground" that everyone on the program agrees to work from.

After your design passes its CDR, it will be "baselined." From that point forward, any change to your design, no matter how small, will require a formal Engineering Change Request. You will hear phrases like "the cost baseline," "the schedule baseline," and "the technical baseline."

The Language of Security

This is the most unique and often intimidating part of the industry. The language of security is not optional; it is a set of rules and concepts that govern your daily professional life. Understanding these terms is non-negotiable for anyone who wants to work on the fascinating, mission-critical projects at the heart of the defense and aerospace world. This is the vocabulary of trust.

Security Clearance

This is the official government determination, made after a thorough background investigation, that an individual is trustworthy and can be granted access to classified national security information. It is the "key to the kingdom." Without the appropriate level of security clearance, you are not legally allowed to work on a classified program. Your job offer will almost always be "contingent" upon your ability to obtain and maintain a clearance.

This will be one of the very first things you do after accepting a job offer. The company's security office will initiate the process on your behalf, which begins with you filling out the SF-86.

Confidential, Secret, Top Secret (TS)

These are the three main levels of classification for national security information, based on the level of damage that would be caused by its unauthorized disclosure. The clearance level required for your job determines the depth of your background investigation and the level of responsibility you will have. Most engineering roles require a Secret clearance. Roles on the most sensitive programs will require a Top-Secret (TS) clearance, which involves a much more comprehensive investigation.

The job description will explicitly state the required clearance level (e.g., "Ability to obtain a Secret Clearance").

SCI (Sensitive Compartmented Information)

A SCI is not a higher level of clearance, but a type of access for specific, highly sensitive intelligence programs, granted on a strict "need-to-know" basis. SCI access is used to protect the nation's most sensitive intelligence sources and methods. It creates "compartments" of information, ensuring that even someone with a Top-Secret clearance cannot access the data unless they have a specific need to know.

You will often see a job posting that requires a "TS/SCI" clearance. This means the role requires both a Top-Secret clearance and eligibility for SCI access. Gaining SCI access often requires a separate, more focused investigation and a polygraph.

SAP (Special Access Program)

SAPs are highly classified "black programs" with security measures and access restrictions that exceed even those of standard SCI. These are established to protect the nation's most advanced and groundbreaking technologies; things like a new stealth aircraft or a next-generation satellite system. The very existence of a SAP can be a secret.

You will likely not know you are applying for a role in a SAP until late in the interview process. Being "read into" a SAP is a significant step that requires the highest levels of trust.

SF-86 (Standard Form 86)

The SF-86 is the lengthy and incredibly detailed questionnaire that you must fill out to begin the background investigation for a security clearance. This is the foundational document for your entire investigation. Your honesty and accuracy on this form are critical. Any deliberate omission or falsification is a federal crime.

You will be given a link to the online e-QIP portal to fill out this form after you accept your job offer. As we will discuss in a later chapter, preparing the information for this form in advance is a critical step.

Adjudication

This is the final step in the clearance process, where a trained government adjudicator reviews all the information from your investigation and makes the formal decision to grant or deny your clearance. The adjudicator is the ultimate decision-maker. They use a formal set of 13 Adjudicative Guidelines and the "Whole Person Concept" to make their risk assessment.

This is a process that happens behind the scenes. You will not meet the adjudicator. Their final decision will be communicated to you through your company's security officer.

Interim Clearance

An Interim Clearance is a temporary clearance that may be granted after the initial checks on your SF-86 are passed, but before the full investigation is complete. An interim clearance allows you to start working on some classified material while the full, lengthy investigation is still ongoing, which can prevent long delays in starting your job.

Your company's security officer will notify you if you have been granted an interim clearance, often a few weeks or months after you submit your SF-86.

SCIF (Sensitive Compartmented Information Facility)

Pronounced "skiff," a SCIF is a secure, often windowless room or an entire facility that has been specially constructed and accredited by the government to handle SCI material. This is the physical embodiment of information security. SCIFs have special locks, alarms, and soundproofing to prevent any classified information from leaving the room.

If you work on a TS/SCI program, you will spend a significant portion of your workday inside a SCIF. You will not be allowed to bring in any personal electronic devices, such as your cell phone or smartwatch.

"Need-to-Know"

This is the fundamental principle of security. It means that you are only granted access to the specific classified information that you have a direct, mission-related need to know in order to do your job. This principle is what creates the "compartments" within the classified world. Just because you and a colleague both have a Top Secret clearance does not mean you are allowed to discuss your classified work with each other.

This will be a constant part of your daily professional life. You will be trained to never discuss your work with anyone who does not have both the appropriate clearance level *and* the specific need-to-know for your program.

ITAR (International Traffic in Arms Regulations)

ITAR is the set of U.S. government regulations that control the export of defense-related technology, services, and information. ITAR is what governs the handling of "sensitive but unclassified" information. It is a legal framework that is designed to prevent critical defense technology from falling into the wrong hands. A violation of ITAR can have severe legal consequences for both you and your company.

From your very first day, you will be trained on ITAR compliance. You will learn that you cannot simply email a technical drawing to a foreign supplier or discuss technical details with a non-U.S. person without a formal export license. It is a fundamental rule of the road for every single person in the industry.

From Vocabulary to Fluency

We have now journeyed through the complex and often intimidating lexicon of the defense and aerospace industry. The sheer volume of terms and acronyms in this chapter can seem overwhelming, like trying to drink from a firehose. The most important piece of advice is this: do not try to memorize them all at once. This chapter, as well as the more exhaustive glossary, is a reference. It is your foundational dictionary, the guidebook you can return to again and again throughout your early career.

Your goal is not rote memorization, but gradual immersion. In your first few months on the job, keep this chapter handy. When you hear a term you don't understand in a meeting or read it in a formal document, look it up. The act of looking it up in the context of a real-world problem will cement its meaning in your mind far more effectively than any flashcard. Ask your mentor or a trusted teammate, "I heard the term 'TRR' in the meeting today. Can you explain what that means in the context of our project?" This shows curiosity and a drive to learn.

As you progress in your career, this vocabulary will become second nature. You will find yourself moving from simply knowing the definitions to achieving true fluency. You will stop translating the terms in your head and start *thinking* in the language of the industry. You will begin to instinctively think about your project in terms of its requirements, its risks, its margins, and its place in the acquisition lifecycle.

This fluency is a powerful and unmistakable marker of your professional growth. It is a clear sign that you have transitioned from a student of engineering, who thinks in terms of homework problems and grades, to a true professional in the defense and aerospace industry, who thinks in terms of systems, processes, and mission success. It is the language of your new tribe and mastering it is a critical step on your journey to becoming a leader within it.

Part 2

The Arenas of Innovation

Chapter 5

A Day in the Life

You've learned about the major companies, key terminology, and the grand, mission-driven nature of the work. But what will you *actually be doing* on a Tuesday morning in October? What is the real, day-to-day texture of a junior engineer's job at a large defense contractor, after the onboarding is over and the real work begins?

The reality of the job is rarely like the montage in a movie. It is not a continuous series of dramatic, last-minute breakthroughs. Nor is it a monotonous, bureaucratic grind. The truth is a fascinating and challenging middle ground, a unique rhythm of deep, focused technical work punctuated by intense collaboration and governed by a structured, disciplined process. Understanding this rhythm is key to thriving in this environment and finding deep satisfaction in your work. This chapter is your immersive guide to that world.

The Two Modes of Engineering Thought

Your time as a defense engineer will be split between two primary modes of working, and learning to master both the rhythm and the transition between them is a critical skill for success. Think of them as the two necessary halves of an engineer's brain: the analytical, focused creator and the communicative, integrated teammate.

1. Deep Work

This is the classic engineering you've pictured since you first decided on your major. It is the sacred, uninterrupted time when you are alone with a difficult technical problem. It is the "headphones on," multi-hour block of pure concentration where you apply the foundational principles of your discipline to create something new or to understand something complex. This is where the tangible engineering gets done. The output of your deep work is the analysis report, the finished CAD model, the debugged block of code, or the processed test data. It is the concrete evidence of your technical contribution.

To excel in deep work, you must become a master of your digital toolkit. Your primary interface with the engineering world will be through a suite of powerful software that turns theory into reality. For a mechanical engineer, your world might revolve around a CAD (Computer-Aided Design) package like Creo or SOLIDWORKS, where you sculpt complex 3D assemblies, and FEA (Finite Element Analysis) software like ANSYS, where you apply virtual forces to see if your design will break. For an electrical engineer, it might be a suite like Altium Designer for laying out a complex Printed Circuit Board (PCB), and a tool like SPICE for simulating its performance. For almost every discipline, the universal language of technical computing is MATLAB/Simulink, the environment where you will model systems, analyze test data, and develop new algorithms. Your proficiency with these tools is your craftsmanship.

This mode of work is deeply satisfying but also intensely demanding. It requires patience, discipline, and a methodical approach. It is the patient process of building a complex simulation model step-by-step, knowing that a single incorrect assumption or a misplaced boundary condition can invalidate hours of work. It is the meticulous act of documenting your code so that another engineer can understand your thought process five years from now. It is the frustration of watching a simulation fail at 99%, and the deep, quiet satisfaction of finally finding the error and achieving a valid solution late in the day. Deep work is the bedrock of your credibility as an engineer. It is where you prove your technical competence, and the quality of your output in this mode will be the foundation of your professional reputation.

2. Collaborative Work

If deep work is about creating the individual pieces of the puzzle, collaborative work is the art of ensuring they all fit together perfectly. On a program with thousands of engineers, no single person has the full picture. A decision made by the thermal team in one building directly impacts the structural, electrical, and software teams in others. Collaborative work is the human network that connects all the individual nodes of deep work into a coherent, functional whole. It consists of the meetings, reviews, and spontaneous discussions that prevent catastrophic errors and ensure the final product is more than just a collection of brilliant but incompatible parts.

This is the "headphones off" time, and it takes many forms, each with its own purpose:

As we first discussed in the introduction, The Formal Design Review (PDR/CDR) are not academic exercises; they are high-stakes events shaping the development of multi-million-dollar, critical national assets where your team presents its design to senior leadership and the customer. The atmosphere is formal, the questions are probing, and the goal is to prove your design is ready for the next phase.

The Working Group (TIM/ICWG) are the smaller, roll-up-your-sleeves meetings focused on solving a specific problem. A Technical Interchange Meeting (TIM) might be called to resolve a conflict between two teams. An Interface Control Working Group (ICWG) is where the precise details of how two systems connect are negotiated and formally documented.

The Peer Review is a cornerstone of the culture. Before any analysis or design is finalized, it is presented to your immediate teammates for review. This is not a test; it is a collaborative effort to find errors, question assumptions, and make the final product better. It requires humility and a thick skin, and it is one of the primary ways young engineers learn from seasoned experts.

The Whiteboard Session is the most creative and spontaneous form of collaboration. It's two or three engineers standing at a whiteboard, sketching out ideas, debating approaches, and trying to solve a problem that has everyone stumped. It is often in these informal sessions that the most elegant "aha" moments occur.

To thrive in collaborative work, you must shift your mindset from "I" to "we." Your goal is not just to prove your solution is correct, but to understand how your solution affects the entire system. Success in this mode is measured by the success of the team and the program. It is in these collaborative forums that you will build your reputation as a trusted teammate and a future leader.

The Rhythm of the Program: A Case Study from Concept to CDR

Let's make this real. We will follow a fictional junior engineer, "Alex," through the lifecycle of a single, critical task. Alex is a mechanical engineer, one year into their job, working on a program to develop a new unmanned aerial vehicle (UAV). Alex's responsibility is the design and analysis of the housing for a critical avionics box. The program is approaching its Critical Design Review (CDR) in one month.

Phase 1: The "Hurry Up" Phase

This is where the magic happens. The entire program is in a "hurry up" mode to finalize, analyze, and document their designs before the major review. The atmosphere in the engineering bay is a low hum of focused intensity. Whiteboards are covered in complex equations and interface diagrams. The energy is palpable. For Alex, this month is a dynamic and demanding blend of deep, solitary work and frantic, crucial collaboration. The goal is simple: get the design right and prove it on paper.

The majority of Alex's time is spent in deep, focused work at their workstation. The task begins in a CAD (Computer-Aided Design) package like Creo. Alex isn't starting from scratch; they are working within a massive top-level assembly of the entire UAV. They can see the surrounding structural ribs, the nearby wiring harnesses, and the hydraulic lines they must avoid. Their job is to design a lightweight aluminum housing that fits perfectly within this digital jigsaw puzzle. It's a delicate balancing act, a constant series of trade-offs. Making a wall thicker for strength adds weight, which the systems team will fight. Adding a mounting hole requires coordination with the structures team. Every design choice has a ripple effect.

Once the initial geometry is created, the work transitions to FEA (Finite Element Analysis) software like ANSYS. Alex methodically builds a simulation model, applying virtual loads that represent the brutal reality the UAV will face: the G-forces of a tight turn, the random vibration of turbulent air, the massive shock of a carrier landing. They spend days applying these loads, analyzing the resulting stress concentrations in the digital model, and iteratively tweaking the design—adding a fillet here, thickening a rib there, changing a material spec—all to ensure the housing will survive without breaking, and without being needlessly heavy.

But Alex isn't working in a vacuum. The deep work is constantly punctuated by vital collaboration. They have a scheduled Technical Interchange Meeting (TIM) with the electrical engineering team to get the critical inputs for the structural and thermal models such as the final mass, dimensions, and heat output of the circuit boards that will go inside the housing. They have a spontaneous whiteboard session with a senior structural analyst, a SME with 25 years of experience, who gives them much-needed advice on how to design the mounting features to avoid a known vibration amplification problem. This 30-minute conversation saves Alex what might have been a week of fruitless trial and error.

Finally, before the design can be "baselined," Alex must present their work at an internal peer review. In a conference room with five other mechanical engineers from their team, Alex walks through the design, the loads, and the analysis results. The feedback is direct, technical, and invaluable. One engineer points out a potential manufacturability issue with a thin wall that would be difficult to machine. Another questions an assumption in the thermal model. This is not a test; it is a collaborative process of finding and fixing errors *before* they get to the customer. Alex takes the feedback, makes the necessary refinements, and finalizes the analysis. The design is ready. With the design baselined and the analysis complete, the focus shifts from creation to justification. The team must now prove their work to the customer in the most demanding forum of all.

Phase 2: The Critical Design Review

The day of the Critical Design Review (CDR) arrives. This is not just a meeting; it is a formal, high-stakes event that can last for days. The main conference room is packed. In the front row sit the customer representatives from the DoD: uniformed officers and senior civilian engineers. Behind them are the program's top leaders: the Program Manager, the Chief Engineer, and lead engineers from every team. The atmosphere is a mixture of intense scrutiny and professional respect. The goal of this review is for the company to prove to the customer, with overwhelming evidence, that the design is complete, mature, and ready for the immense expense of manufacturing.

The lead engineer for Alex's team, the Integrated Product Team (IPT) Lead, is at the podium. They are responsible for presenting the entire avionics suite. The presentation is a massive PowerPoint deck, hundreds of slides long, representing months of work by dozens of engineers. Alex's contribution is a small but critical part of this larger whole. Their name is on the bottom of the slides detailing the housing's structural and thermal analysis. They sit in the audience, heart pounding, waiting for their section to come up.

The lead engineer clicks to a slide showing a 3D model of Alex's housing, with a color-coded FEA plot showing the stress distribution under the highest load case. The lead presents the key findings: "As you can see, the maximum stress in the housing is well below the material allowable, with a positive margin of safety, and the peak temperature of the processor remains within its operational limits."

After the lead finishes the section, the customer's Chief Engineer, a civilian with decades of experience, leans into the microphone. "That's a good analysis," he begins, "but your vibration load case seems standard. This platform will have a significant acoustic load profile during takeoff. What was your fatigue analysis for the mounting lugs under that combined acoustic and random vibe environment?"

This is the moment of truth. An unprepared team might falter. But Alex's lead engineer is ready. "Excellent question, sir," he says, without missing a beat. "If you'll turn to page 47 of your briefing packet, and I'll bring up Backup Slide 21." On the main screen, a new slide appears, showing the detailed fatigue analysis that Alex had painstakingly completed two weeks prior, specifically to address this exact concern after a senior mentor had suggested it. The lead walks through the analysis, showing that the design has a calculated fatigue life of over four times the required service life. The customer's chief engineer nods, satisfied. "Good work," he says. The review moves on. Alex feels a profound sense of relief and professional pride. In that single moment, the countless late nights, the frustrating dead ends, and the meticulous peer reviews all became worth it. All those hours of meticulous deep work, the peer reviews, and the advice from a mentor were for that one moment of validation.

Phase 3: The "Wait" and the "Real Work"

The CDR was a success. The program has received approval to proceed with manufacturing the first prototypes. The intense, adrenaline-fueled "hurry up" phase is over. Now, a different rhythm begins: the "wait." The procurement team begins the long, formal process of placing orders for the thousands of custom-machined parts, a process that can take weeks or months. For Alex, this is not downtime. The nature of the work simply shifts from creation to integration and future-planning. The focus moves from deep work to almost exclusively collaborative work.

Alex's calendar, once cleared for focused design time, now fills up with meetings, but these are not pointless status updates; they are where the next stage of engineering happens. Alex is a key participant in an Interface Control Working Group (ICWG). The goal of this meeting is to finalize the Interface Control Document (ICD), a legally binding contract between Alex's team and the wiring harness team. They go through the document line by line, formally agreeing on the exact connector part numbers, the function of every single pin, the shielding requirements, and the precise physical location of the connector on the housing. This meticulous, and sometimes tedious, work is what ensures the thousands of pieces of the final UAV will actually plug together and function correctly.

Later in the week, Alex participates in a "lessons-learned" meeting with their IPT. The team discusses what went right and what went wrong during the CDR push. They identify a flaw in their peer review process that allowed a minor design issue to slip through, and they brainstorm a way to fix it for the next review cycle. This is the process of continuous improvement in action.

Finally, Alex is tasked with a forward-looking project. With the current design frozen, the lead engineer asks Alex to begin a trade study for a future, lighter-weight version of the housing. This is a chance for deep work again, but of a different kind. Alex spends their time researching new aluminum alloys, exploring the possibility of using a composite material, and doing a high-level analysis of the potential weight savings versus the increased manufacturing cost. This work will form the basis of the next design spiral, months or even years in the future. By understanding that the "wait" is not downtime but a different kind of "up-time," Alex transforms these periods from frustrating lulls into some of the most productive and career-enhancing phases of their work. This is the true rhythm of a defense engineer.

Alex's journey is a microcosm of the engineering life you are about to enter. It is a dance between the quiet intensity of deep work and the dynamic challenges of collaborative problem-solving. It is a cycle of frantic, deadline-driven sprints followed by periods of urgent integration and strategic, forward-looking planning. Understanding and mastering this rhythm of knowing when to put the headphones on and focus, and when to take them off and engage is the fundamental, unwritten skill of the effective defense engineer.

Chapter 6

The Arenas of Innovation

For all of human history, conflict has been defined by geography. The fundamental arenas of competition were dictated by the physical realities of the planet itself: the vast plains, dense jungles, and rugged mountains of the Land; the dark, unforgiving oceans of the Sea; and the boundless complexity of the Air. These three domains were distinct worlds. An army mastered the science of ground warfare, a navy mastered the art of seamanship, and an air force, when it arrived, mastered the new physics of flight. They were separate services, each populated by engineers and tacticians who were deep specialists in their own environment.

The latter half of the 20th century, however, ripped open the fabric of this reality. Two new, invisible arenas emerged that would fundamentally reshape the nature of conflict and, consequently, the nature of engineering. The first was the harsh, limitless vacuum of Space, a domain of ultimate high ground where orbital mechanics and cosmic radiation were the new laws of physics. The second was the man-made, instantaneous world of Cyberspace, an electronic battlefield where invisible wars could be fought at the speed of light.

Today, no domain stands alone. They are deeply, inextricably intertwined. A soldier on the ground is utterly dependent on a precise timing signal from a GPS satellite orbiting 12,000 miles above the Earth. A pilot in the air is protected by an electronic warfare system that is actively fighting a battle in the electromagnetic spectrum. A ship at sea is no longer an isolated fortress, but a powerful, data-hungry node in a global network. This new reality is known as Multi-Domain Operations, and it is the core strategic concept that drives modern defense engineering.

Understanding these five domains, Air, Land, Sea, Space, and Cyber, is therefore the most powerful tool you have for navigating your career. It allows you to move beyond a generic search for a "mechanical engineering job" and begin a targeted, passionate pursuit of a career that truly excites you. It is the difference between telling a recruiter, "I'm looking for an internship," and telling them, "I'm passionate about the unique structural and thermal challenges of hypersonic flight, and I'm eager to contribute to the Air Domain." This chapter is your strategic guide to this new, multi-domain world.

The End of the Monoliths

For much of the 20th century, military dominance in one domain could often guarantee victory. A superior navy could control the seas, and a superior air force could command the skies. These were powerful, but largely separate, pillars of national power. That era is over. The past two decades have seen a fundamental shift in the character of warfare, driven by two key trends that have made the old, siloed approach obsolete.

First is the proliferation of advanced technology. For decades, the United States held a near-monopoly on technologies like precision guidance, stealth, and advanced sensor networks. Today, those technologies have spread. Potential adversaries can now field sophisticated anti-ship missiles that threaten navies, advanced integrated air defense systems that challenge air forces, and GPS jammers and anti-satellite weapons that contest the space and cyber domains. The "sanctuary" is gone. No single domain is safe, and no single domain can win alone.

Second is the primacy of data. The battlefield is no longer just a physical space; it is an information space. The victor is not necessarily the one with the biggest platform, but the one who can gather data from every possible source, fuse it into a coherent picture of reality, and act on it faster than the enemy. A single platform, no matter how capable, has a limited view. A networked force of platforms, however, becomes a single, distributed sensor, able to see the battlefield from the seabed to low Earth orbit, creating an overwhelming information advantage.

Multi-Domain Operations is the answer to this new reality. It is the recognition that a single, integrated "kill web" is vastly more lethal, resilient, and effective than the sum of its individual parts. This creates the primary engineering challenge of our time: designing not just great platforms, but the resilient networks, shared data standards, and intelligent algorithms that connect them into a single, cohesive whole.

The Multi-Domain Battlefield in Action

To understand this concept, let's move beyond the abstract. Consider a single, critical moment for a soldier on the ground, tasked with identifying a hostile target hidden in a dense urban environment. She is operating in the Land Domain, but her success is utterly dependent on the seamless integration of the other four.

The signal she receives comes from Space. A precise GPS timing and navigation signal, originating from a satellite constellation orbiting 12,000 miles overhead, pinpoints her exact location on a digital map. That same satellite network provides the beyond-line-of-sight communications link that keeps her connected to her command.

The data flowing over that link is protected by Cyberspace. The communications are encrypted, and the network itself is hardened against enemy jamming and intrusion attempts. A team of cyber-protection engineers, hundreds of miles away, are actively defending the network she is relying on.

Her eyes in the sky are from the Air Domain. An unmanned aerial vehicle (UAV) loiters silently thousands of feet above, its powerful electro-optical sensor streaming full-motion video directly to the tablet in her hands, allowing her to see around the next corner without exposing herself to danger.

The final effect is delivered from the Sea Domain. When she confirms the target, that targeting data is passed through the network to a Navy destroyer operating a hundred miles off the coast, which can then launch a precision munition to neutralize the threat.

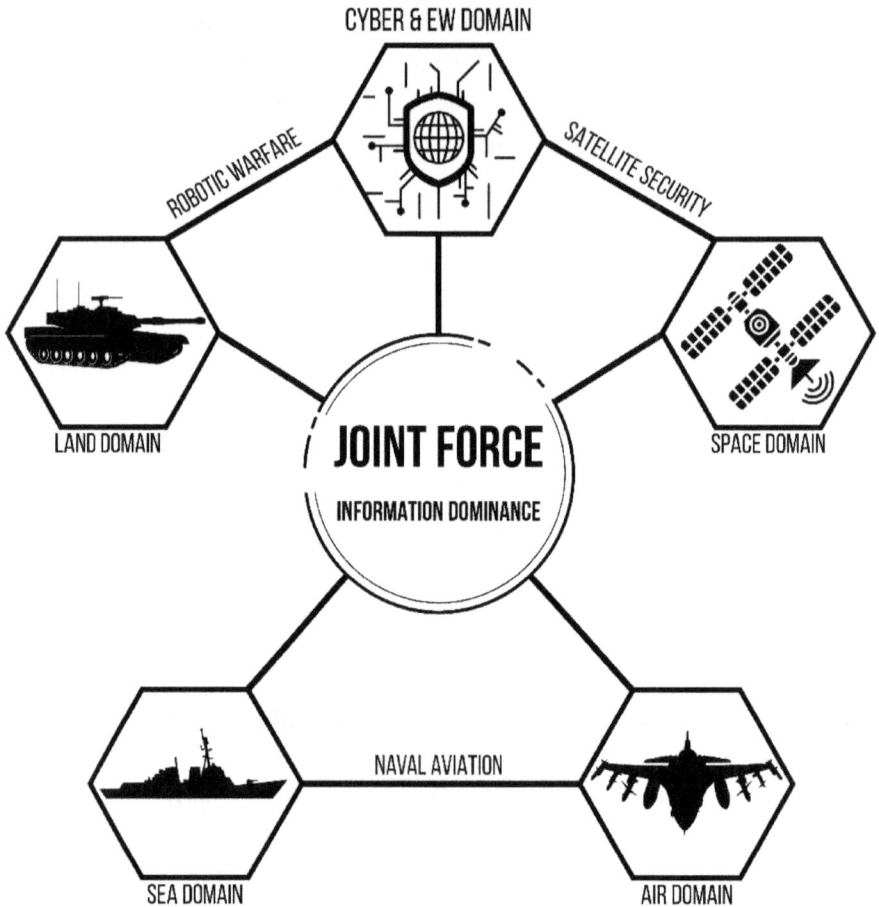

Figure 3: *To achieve information dominance, the modern military operates as a single, interconnected "Joint Force," not as separate services. The Joint Force integrates capabilities across the five primary domains: Air, Land, Sea, Space, and Cyber. The most complex and innovative engineering challenges are often found in the "seams" between these domains, such as Naval Aviation, Robot Warfare, and Satellite Security.*

The soldier on the ground never sees the satellite, the cyber operator, the drone pilot, or the sailor. Yet, she is the focal point of a vast, interconnected system of systems. This is Multi-Domain Operations. It is the core strategic concept that drives modern defense engineering, and it creates an insatiable demand for engineers who can think, design, and build across these once-separate worlds.

The Seams Between the Worlds

While we will explore each domain in detail, it is often in the seams, the fascinating, complex, and often brutal intersections between them, where the most difficult and rewarding engineering challenges lie. These overlaps are where the future is being invented, and they are hungry for engineers who can think across traditional disciplines, who are fluent in more than one technical language. The most valuable engineers of the 21st century are those who can stand with a foot in two different worlds.

Consider the violent and beautiful seam between the Air and Sea domains. This is the world of naval aviation, arguably one of the most demanding engineering environments on Earth. The challenge is not just to build a jet, but to build a jet that can withstand a controlled crash onto a pitching, rolling runway just a few hundred feet long. It must be launched by a catapult that accelerates it from zero to 170 miles per hour in under three seconds, placing unimaginable stress on its airframe. It must then fold its wings to fit into a crowded hangar bay and operate for decades in a corrosive saltwater environment that is actively trying to eat it alive. This seam is a harsh intersection of aerodynamics, structural mechanics, materials science, and corrosion control.

Or consider the invisible, high-stakes seam between the Space and Cyber domains. A modern satellite is not just a piece of hardware; it is a sophisticated, software-driven, networked computer hurtling through a radioactive vacuum at 17,500 miles per hour. How do you protect that computer, which is a priceless national asset, from being disabled or hijacked by a malicious signal sent from a hacker on the other side of the world? How do you write code that is so utterly reliable that it can be trusted to operate for 20 years without a reboot, all while being constantly bombarded by solar radiation that can flip a digital bit from a zero to a one in an instant? This is a world that demands engineers who are fluent in both orbital mechanics and cybersecurity, a rare and incredibly valuable combination.

The seam between the Land and Cyber domains is where the science fiction of robotics and artificial intelligence is becoming a battlefield reality. The challenge is not just to build a robotic vehicle, but to build one that can navigate a cluttered, hostile, and GPS-denied environment, distinguishing between enemy combatants, friendly forces, and non-combatants. It must be able to collaborate with human soldiers as a true teammate, all while its sensors and communication links are under constant electronic and cyber-attack. This is the new frontier for AI and machine learning, a world that needs mechanical engineers who understand software and software engineers who understand the physical world.

A Deeper Look at the Seams

Let's dissect these seams to see where specific engineering disciplines are most in demand. These are the nexus points where the most complex and rewarding problems are being solved.

The Air-Sea Seam

The Mission Challenge: Operating high-performance aircraft from the violent, corrosive, and space-constrained environment of a pitching aircraft carrier.

Engineering Problem Examples: High-G structural loads from catapult launches and arrested landings; advanced materials science for corrosion control; complex hydrodynamics of the ship-air interface; high-density systems integration.

Engineers in High Demand: Mechanical, Aerospace, Materials Science, Electrical (for example, launch and recovery systems).

Keywords for You: Structural Dynamics, Fatigue Analysis, Corrosion Control, Materials Science, Systems Integration.

The Land-Cyber Seam

The Mission Challenge: Enabling manned and unmanned ground vehicles to collaborate on a chaotic, GPS-denied battlefield while under direct cyber and electronic attack.

Engineering Problem Examples: Autonomous navigation in unstructured environments; computer vision and sensor fusion for target identification; secure, ad-hoc mobile networking; human-machine interface design.

Engineers in High Demand: Computer Science (AI/ML), Computer Engineering, Electrical, Mechanical (for example, robotics and mechatronics).

Keywords for You: Robotics, ROS (Robot Operating System), Computer Vision, AI/Machine Learning, Network Security, Human Factors.

The Space-Cyber Seam

The Mission Challenge: Protecting priceless national assets in orbit, which are essentially high-performance computers in a radioactive vacuum, from being hacked, jammed, or disabled.

Engineering Problem Examples: Writing ultra-reliable, fault-tolerant flight software; designing radiation-hardened electronics; securing satellite command and data links with advanced cryptography; building resilient network architectures.

Engineers in High Demand: Computer Science, Computer Engineering, Electrical, Aerospace (for example, electronic warfare).

Keywords for You: Embedded Systems, Fault-Tolerant Computing, Cybersecurity, Cryptography, Radiation Hardening, Astrodynamics.

Finding Your Interdisciplinary Edge

Ask yourself:

Are you a mechanical engineer with a passion for coding and software? The world of mechatronics, robotics, and control systems lives at the seams of Land, Sea, Air, and Cyber. Are you an electrical engineer fascinated by the physics of space? The world of satellite communications, RF payloads, and radiation-hardened electronics needs your expertise at the seam of Space and Cyber. Are you a computer scientist who loves tangible hardware and the physical world? The future of autonomous navigation, computer vision, and manned-unmanned teaming requires your skills to leave the server farm and operate at the tactical edge.

Finding your passion within one of these seams makes you an incredibly valuable and compelling candidate. It allows you to build a unique narrative that is far more powerful than a simple declaration of your major.

Using the Domains to Craft Your Career Story

The five-domain model is your first piece of actionable intelligence. It provides a strategic lens through which you can analyze the entire defense industry, transforming your job hunt from a random walk into a precise campaign. This framework is the key to telling a compelling story, targeting your applications effectively, and asking the intelligent questions that will make you stand out.

1. Go Beyond a Generic Title

This framework allows you to define yourself with precision and passion. It elevates your introduction from a simple statement to a compelling narrative.

The Generic Student: "Hi, I'm a junior mechanical engineering student and I'm looking for an internship."

The Strategic Candidate: "Hi, I'm a junior mechanical engineer, and I'm passionate about the unique structural and thermal challenges of hypersonic flight. I'm excited to find a role in the Air Domain where I can contribute to that mission."

The second introduction is in a completely different league. It tells the recruiter not just what you are, but *who you want to become*. It demonstrates focus, passion, and an understanding of their world.

2. Target Your Job Search with Precision

The defense industry is not a monolith. A company like Northrop Grumman may have a facility in Baltimore that specializes in radars (Air/Cyber Domain) and another in Redondo Beach that builds satellites (Space Domain). Using the domains as a filter, you can intelligently target your search to the specific companies, and even the specific *locations*, that are doing the work that excites you. This prevents you from wasting time applying to roles that are a poor fit and allows you to concentrate your efforts where they will have the most impact.

3. Win the Interview by Speaking Their Language

Walking into an interview with this framework in mind is a superpower. It allows you to ask more intelligent questions and connect your own experience to their mission.

A Generic Question: "What kind of projects do interns work on?"

A Domain-Aware Question: "I was fascinated by the discussion of the 'seam' between the Space and Cyber domains. Could you talk about the specific challenges your team faces in securing satellite data links?"

This level of questioning immediately elevates you from "student" to "potential colleague." It proves you have done your homework and are already thinking at a strategic level. It is how you find your arena in this complex and fascinating multi-domain world.

How to Become an Expert in a Domain

Your passion for a domain will be most convincing when it's backed by specific knowledge. You don't need a security clearance to become well-informed. The strategic conversations about the future of these domains are happening in public. Investing a few hours a week in reading can give you the knowledge of an industry insider.

Your Weekly Briefing:

Primary News Sources: Websites like Breaking Defense, Defense News, and C4ISRNET are the daily papers of the industry. They report on new contracts, emerging technologies, and key policy decisions.

Deeper Dives: For the Air and Space domains, Aviation Week & Space Technology is the gold standard. For naval issues, the U.S. Naval Institute's Proceedings is essential reading.

Think Tank Analysis: Organizations like the Center for Strategic and International Studies (CSIS) and the RAND Corporation publish deep, insightful analysis on the strategic and technological trends shaping the future of the military. Their reports are often the source of the ideas that become official policy years later.

Follow the Innovators: Follow organizations like DARPA, the Air Force Research Laboratory (AFRL), and the Office of Naval Research (ONR) on social media. They are constantly sharing unclassified information about the cutting-edge research they are funding.

Your Journey Through the Arenas

This section of the book is your strategic briefing, a deep dive into each of these five arenas. We will move beyond simple descriptions and immerse you in the culture, the challenges, and the iconic programs that define each one. This is your chance to discover which world speaks to you, to find the set of problems that you want to dedicate your career to solving.

First, we will take to the skies in the Air Domain. This is the classic world of aerospace, a domain of speed, stealth, and the absolute limits of flight. We will explore the elegant complexity of a modern fighter jet, the silent menace of a stealth bomber, and the physics-defying challenge of hypersonic flight. This is the arena for those who are captivated by the beauty and the brutality of aerodynamics.

Next, we will come back to earth in the Land Domain. This is the most visceral of the arenas, the world of armor, robotics, and the engineering required to operate in direct contact with the ground. We will examine the challenges of protecting a vehicle from a blast, of designing a robotic system that can navigate a forest, and of building equipment that can survive a lifetime of bone-jarring vibration.

Then, we will go to sea, exploring the immense scale and complexity of the Sea Domain. This is the ultimate systems integration challenge, a world of floating cities and silent, nuclear-powered predators. We will explore the art of building a warship that can fight, survive, and sustain a crew for months at a time, and the deep, dark secrets of submarine acoustics.

From there, we will launch into the final frontier in the Space Domain. This is a world of perfect precision and absolute reliability, where the tyranny of the rocket equation governs every decision. We will explore the challenges of designing satellites that must operate for decades in the harshest environment known, from the GPS constellation that runs our world to the reconnaissance satellites that serve as our eyes in the sky.

Finally, we will log into the invisible battlefield of the Cyber & Electronic Warfare Domain. This is the modern arena where conflicts are increasingly won and lost at the speed of light. We will explore the world of signals, data, and algorithms, and the deep intellectual challenge of controlling the electromagnetic spectrum.

For each domain, we will explore its crown jewel programs, its core engineering challenges, a snapshot of a day in the life, and the key companies and government labs that are the dominant forces in that arena.

As you read through the following chapters, don't just passively absorb the information. Ask yourself: Which of these problems truly sparks my curiosity? Which of these missions resonates with my own sense of purpose? Answering that question is the first step toward crafting a compelling story for your resume, your cover letter, and your interviews. It is how you find your arena in this complex and fascinating multi-domain world.

Chapter 7

The Air Domain

Speed, stealth, and the absolute limits of physics define the Air Domain. Here, in the realm of blistering speed and vanishingly low observability, the most significant engineering breakthroughs have been forged. From breaking the sound barrier to achieving functional invisibility with stealth, the challenges of achieving air superiority have consistently driven technology forward at a breathtaking pace. This is the domain for engineers who are captivated by the physics of flight and the challenge of building machines that operate at the very edge of what is possible.

For a young engineer, this domain offers a chance to put your signature on some of the most advanced and recognizable platforms on Earth. It is the world of sleek fighter jets that can pull 9 G's, silent, continent-spanning bombers, agile, gravity-defying helicopters, and the intelligent, autonomous drones that are the future of air power. But it is so much more than that. The Air Domain is an intricate, layered "system of systems." It is a delicate dance between the platforms that fly, the smart munitions they carry, the network of sensors that watch the battlefield, and the command-and-control systems that act as the "brain" of the entire operation. The work is a constant, high-stakes dialogue between thrust, lift, weight, and drag, but also between data links, sensor resolution, and algorithmic speed. It is a world where every gram of mass and every line of code is meticulously accounted for.

From the Sound Barrier to the Fifth Generation

To understand the engineering challenges of the modern Air Domain, you must first understand the story of how we got here. The history of military aviation is a relentless, multi-generational staircase of technological leaps, where each step was a response to the threats and limitations of the last. For fighter aircraft, this is formally categorized into "generations," and understanding them is key to understanding the modern engineering mindset.

The First and Second Generations (the 1940s-1950s) were about one thing: speed and altitude. This was the era of legendary test pilots like Chuck Yeager, a time when the primary engineering challenge was breaking the sound barrier and building jet engines that could push an aircraft faster and higher than the enemy. The engineering was raw, daring, and focused on pure aerodynamic performance.

The Third and Fourth Generations (the 1960s-1990s), exemplified by iconic aircraft like the F-4 Phantom, F-15 Eagle, and F-16 Fighting Falcon, were about energy and missiles. As the skies filled with supersonic jets, the focus shifted to the ability to out-turn an opponent in a dogfight, maneuverability, and the development of sophisticated radar systems and guided missiles that could engage a target from beyond visual range. This is when the deep integration of advanced avionics and control systems began to define a fighter's lethality.

The Fifth Generation, which is the reality of today, is defined by three revolutionary concepts: stealth, sensor fusion, and networking. The engineers who designed the F-22 Raptor and the F-35 Lightning II understood that the next great leap was not about being faster or more maneuverable, but about being smarter and invisible. The core engineering challenge shifted from physics to information. The most lethal aircraft is the one that can see its opponent long before it is seen itself, share that information seamlessly with the rest of the force, and make the first move. This is why a modern program like the F-35 is often described as a "flying supercomputer."

As an engineer entering the Air Domain today, you are stepping into this legacy. The work you do is part of the race to define the Sixth Generation, which will almost certainly be defined by artificial intelligence, optional manning, and the ability to command swarms of autonomous drones. The challenges we will discuss in this chapter (the obsession with weight, the complexity of software, the science of advanced materials) are all born from this relentless, seventy-year pursuit of the technological edge.

The Crown Jewel Programs

The work in this domain is not just about building aircraft; it's about creating a complex, integrated "kill web" that can control the sky. This web has three primary components: the platforms that provide presence, the munitions that provide lethality, and the command and control systems that provide the intelligence.

1. The Platforms (The "Shooters" and "Eyes")

These are the most visible and iconic parts of the Air Domain, the physical hardware that operates in the atmosphere.

The F-35 Lightning II: Far more than a simple fighter jet, this platform acts as the most advanced and complex flying sensor and data-networking platform ever created. Serving as Lockheed Martin's flagship program, it represents a massive leap in "sensor fusion," where data from its advanced radar, 360-degree infrared sensors, and electronic warfare suite are all combined by powerful software into a single, intuitive picture for the pilot.

The B-21 Raider: The successor to the iconic B-2 Spirit, the B-21 Raider from Northrop Grumman represents the absolute cutting edge of low-observable (stealth) technology. This program is shrouded in secrecy, but its engineering challenges are immense, from developing new radar-absorbent materials to pioneering new manufacturing techniques.

The KC-46 Pegasus: As one of the unsung heroes of the domain, the KC-46 serves as the critical enabler that gives U.S. air power its global reach, allowing fighters and bombers to stay airborne for extended periods. The engineering is a complex blend of commercial aircraft design and military-specific systems.

Unmanned Systems: The work spans the full spectrum of unmanned systems, from the Northrop Grumman Global Hawk, a high-altitude, long-endurance UAV with the wingspan of a Boeing 737, to the iconic

General Atomics Predator/Reaper family, the workhorses of persistent surveillance and strike.

2. The Munitions (The "Arrows")

An aircraft is only as effective as the weapons it carries. The engineering that goes into these munitions is as advanced as the platforms themselves.

AMRAAM (Advanced Medium-Range Air-to-Air Missile): Produced by Raytheon, this is the world's most sophisticated air-to-air missile. The engineering challenges are immense: fitting a powerful rocket motor, a miniature radar seeker, a warhead, and a flight computer into a small, seven-inch-diameter tube that must withstand the severe G-forces of a fighter jet's maneuvers.

JDAM (Joint Direct Attack Munition): The technology that turns a "dumb" bomb into a smart, GPS-guided precision weapon. This program, led by Boeing, revolutionized modern air-to-ground warfare by creating an affordable, all-weather guidance kit. The engineering lies in the robust GPS/INS (Inertial Navigation System) that can guide the weapon with pinpoint accuracy.

3. The Command and Control (The "Brain")

This is the invisible network that connects all the pieces and makes them more than the sum of their parts.

AWACS (Airborne Warning and Control System): This is the iconic Boeing 707 with a giant rotating radar dome on top. It is a flying command post, a "quarterback in the sky" that manages the air battle, detecting enemy aircraft from hundreds of miles away and directing friendly fighters to intercept them.

JADC2 (Joint All-Domain Command and Control): This is not a single program, but the massive, ongoing effort to create a "combat cloud." The goal is to build the software, resilient networks, and AI that will connect every sensor to every shooter across all domains (Air, Land, Sea, Space, and Cyber) in real-time. This is the new frontier of C4ISR (Command, Control, Communications, Computers, Intelligence, Surveillance, and Reconnaissance), and it is almost purely a software and network engineering challenge.

4. The Unsung Heroes: Enablers of Air Power

While stealth fighters and continent-spanning bombers capture the imagination, they are only able to perform their mission because of a vast ecosystem of less-famous but equally complex "enabling" aircraft. A significant portion of the most stable and interesting engineering work in the Air Domain is on these vital platforms.

Aerial Refueling ("Flying Gas Stations"): Without tankers, the United States Air Force would be a regional power. Platforms like the KC-46 Pegasus (Boeing) are the lynchpin of global reach. These are not simple transport aircraft; they are complex, flying fuel depots. The engineering is a fascinating blend of commercial aircraft design (the 767 airframe) and unique military systems, chief among them the flying boom. Designing a rigid but maneuverable, 50-foot telescoping boom that can safely connect to another aircraft in mid-air involves incredible challenges in control systems, hydro-mechanical engineering, and remote vision systems.

Military Transport ("Heavy Lifters"): The ability to move troops and equipment anywhere in the world in under 24 hours is a cornerstone of modern strategy, and it rests on the wings of aircraft like the iconic C-130 Hercules (Lockheed Martin) and the massive C-17 Globemaster III (Boeing). The engineering on these platforms is a masterclass in structural efficiency and durability. The challenge is designing an airframe that can carry an M1 Abrams main battle tank, operate from short, unimproved, or even dirt runways, and endure decades of punishing, high-cycle operational use.

Rotary Wing ("Agile Support"): Helicopters and tilt-rotor aircraft operate in a world of staggering mechanical and aerodynamic complexity. Platforms like the H-60 Black Hawk family (Sikorsky/Lockheed Martin) a the AH-64 Apache (Boeing) are less like planes and more like intricate, flying transmissions. The engineering of the rotor systems with their swashplates, blade grips, and vibration-damping systems is one of the most demanding niches in mechanical and aerospace engineering, a constant, high-stakes dance between performance, structural fatigue, and vibration control.

Core Engineering Challenges

Aerodynamics & Control Systems: This remains the fundamental challenge of flight. It involves using Computational Fluid Dynamics (CFD) to optimize an aircraft's shape for performance and designing the sophisticated, often triple-redundant "fly-by-wire" control systems that translate a pilot's commands into stable, responsive flight, especially for inherently unstable stealth platforms.

Structures & Materials: We define this as the discipline of building things that are impossibly light and unbelievably strong. The work is a constant war against weight, driving innovation in advanced composite materials, additively manufactured (3D printed) metal alloys, and sophisticated Finite Element Analysis (FEA) to ensure the airframe can survive the harsh G-forces, vibrations, and temperatures of flight.

Stealth (Low Observability): We define this as the discipline of building things. It involves shaping an aircraft with complex curves and angles to deflect radar waves, developing deep materials science expertise for radar-absorbent materials (RAM), and meticulously managing thermal and acoustic signatures to make the platform as difficult to detect as possible.

Guidance, Navigation, and Control (GNC): This is the core challenge for missiles, munitions, and autonomous drones. It is a world of complex algorithms, Kalman filters, and the fusion of data from GPS receivers and Inertial Measurement Units (IMUs), all designed to ensure a system can find its target with pinpoint accuracy.

RF and Sensor Engineering: This is the heart of the ISR and C4ISR world. It is the physics of radar, the design of sophisticated electro-optical and infrared cameras, and the engineering of the antennas and receivers that allow platforms like the AWACS to "see" hundreds of miles away.

Software and Network Engineering: This is the glue that holds the entire modern Air Domain together. The challenge is writing millions of lines of ultra-reliable, real-time code for flight-critical systems and designing the resilient, secure data networks that allow all platforms to communicate and collaborate as a single, cohesive force.

Propulsion & High-Temperature Systems (The Heart of the Beast): An aircraft is just a sculpture without a powerplant. The jet engine is arguably the most extreme and unforgiving engineering environment on Earth. The core of a modern fighter engine operates at temperatures hot enough to melt the very metal it's made from. Engineers here are locked in a constant battle to push materials to their absolute limits, inventing exotic single-crystal turbine blades and ceramic matrix composites that can survive these infernal conditions. The work involves a deep mastery of thermodynamics, computational fluid dynamics (CFD) for airflow, and incredible structural analysis to manage the immense rotational stresses, where a single turbine blade spinning at over 10,000 RPM has the kinetic energy of a speeding car. Every ounce of thrust is the product of this violent, controlled explosion, and for a propulsion engineer, there is no more rewarding challenge.

A Day in the Life: Five Perspectives

The experience of an engineer in the Air Domain varies dramatically based on their role and specialty. While they all work toward the same mission, their daily challenges, tools, and environments can feel like different worlds. Let's look at a week for five different junior engineers.

The Structures Analyst:

A mechanical engineer like "Alex" spends the vast majority of their time in the digital world. Their week is focused on a single component: a newly designed wing rib for a next-generation fighter. Using a Finite Element Analysis (FEA) tool like ANSYS or NASTRAN, their mission is to ensure the design can withstand the 9-G forces of a high-speed turn while being as light as possible. They meticulously apply digital loads and boundary conditions to their model, run the analysis, and then spend hours interpreting the color-coded stress plots. They are digital sculptors, iteratively shaving a few grams of weight from a pocket here, thickening a fillet there, all in a relentless quest to optimize the design's strength-to-weight ratio. Their primary interactions are deep, technical peer reviews with other structures experts, where every assumption in their analysis is challenged.

The GNC Engineer:

An aerospace or software engineer's world lives in simulation. Their task is to improve the terminal guidance algorithm for an air-to-air missile. Their primary tool is MATLAB/Simulink. They aren't building hardware; they are architecting logic. Their week is spent tweaking the code for a Kalman filter that fuses data from the missile's radar seeker and its IMU. The goal is to create a navigation solution so robust that it can maintain a lock on a highly maneuverable target even in the presence of enemy jamming. The climax of their week is not a physical test but running a Monte Carlo simulation (a batch of 5,000 virtual launches where every parameter is randomized) to prove with statistical certainty that their new algorithm has improved the missile's probability of kill by a few precious percentage points.

The Systems Engineer:

An aerospace or systems engineer spends their week in meetings, but these meetings are the work. They are the human network that connects all the other disciplines. Their morning might be spent in a Technical Interchange Meeting (TIM) trying to resolve a conflict between the thermal team, who needs to add a cooling vent, and the stealth team, for whom that vent is a catastrophic source of radar reflection. The systems engineer's job is not to be an expert in either field, but to understand the constraints of both, facilitate the trade study, and guide the teams to a compromise that works for the whole aircraft. Their afternoon could be in an Interface Control Working Group (ICWG), meticulously documenting the pin-outs and data formats for a connector that links a sensor from one contractor to a processor from another. Their product is not a part or a piece of code, but clarity, communication, and disciplined process.

The Test Engineer:

An electrical or software engineer might spend their week in a cavernous System Integration Lab (SIL), a massive, windowless room filled with racks upon racks of the real, flight-worthy electronics for the aircraft. The air hums with cooling fans. Their job is to be the "first customer," the person who tries to break the system on the ground so that it never fails in the air. They write scripts that run the system through thousands of simulated missions, testing every line of code and every electronic interface. They are digital detectives. When a test fails, their job is to isolate the problem, digging through gigabytes of diagnostic data to determine if the root cause was a software bug, a hardware failure, or an incorrect interface, and then work with the design teams to find a solution.

The RF Engineer:

An electrical engineer specializing in stealth might spend their week in an almost sacred space: the anechoic chamber. The walls, ceiling, and floor are covered in massive, blue, pyramid-shaped foam cones designed to absorb all electromagnetic energy, creating a perfectly silent RF environment. Dressed in a special anti-static "bunny suit," their task is to test a new radar-absorbent paint sample. Using a network analyzer, they meticulously measure the material's reflectivity across a wide spectrum of radar frequencies, comparing the physical test data to the predictions from the electromagnetic simulation models. Their work is a quiet, precise, and highly empirical search for the materials that will make the next generation of aircraft effectively invisible.

What Makes The Air Domain Unique?

Every engineering domain is ruled by a brutal, unforgiving tyrant. In the Sea Domain, it is crushing pressure and corrosive salt. In the Land Domain, it is shock and vibration. In the Air Domain, the tyrant is gravity. Every single aspect of the culture, every process, and every obsession of an aerospace engineer is a direct consequence of this relentless, career-long war against weight. Understanding these cultural drivers is the key to adapting and thriving.

Weight is the Enemy, Always: This is the first and most important commandment of aerospace engineering. It is not just one of several competing constraints; it is the primary one that governs almost every decision. A land systems engineer might add a pound of steel for reliability with little consequence, but for an aerospace engineer, that pound has a cascading and punishing effect due to the "tyranny of the rocket equation." In practice, this obsession with mass is not abstract; it will define your daily work.

Your first major task may be a "weight scrub," where you are asked to go through every component down to the bolts and washers to find a lighter alternative. In every design review, you will be expected to present your "mass properties" and defend your part's weight, down to the gram, against intense scrutiny. You will quickly learn that a brilliant design that is overweight is not a brilliant design at all.

The Primacy of Configurations: In the Air Domain, the external shape of the platform, or the "Outer Mold Line" (OML), is treated as an almost sacred geometry. This shape is the product of thousands of hours of aerodynamic and stealth optimization. This strict adherence to the OML directly translates into the rigorous process of Configuration Management (CM). As an engineer, you will discover that you cannot add even a tiny external sensor pod without triggering a major review. You may be tasked with performing a CFD analysis to prove a change won't cause unexpected turbulence, or work with stealth experts to validate that a new feature doesn't create a new radar reflection. You will learn a fundamental lesson of aerospace: you don't truly "own" the outside surface of your design; it is owned by the entire program.

A Culture of Deep Specialization: Because the physics of flight, propulsion, and stealth are so incredibly complex, the Air Domain fosters a culture of deep, narrow specialization. The hero of this culture is the Subject Matter Expert (SME), the engineer who has spent decades becoming a world-class authority in a single niche. For your early career, this means your primary goal is to forge a deep, vertical expertise. You will not be a generalist; you will strive to become the "wing rib FEA expert" or the "landing gear kinematics specialist." Your value, and the beginning of your reputation, will be measured not by the breadth of your knowledge, but by its depth.

The "Flight-Critical" Standard of Rigor: For any system whose failure could lead to the loss of the aircraft and its crew, the engineering discipline rises to a level of paranoia that borders on religion. This concept of "flight criticality" creates a two-tiered system of rigor. If you find yourself assigned to a flight-critical component, like a flight control actuator, you will immediately encounter this elevated world of scrutiny. Your analysis will require more validation, your hardware will require triple or quadruple redundancy, and your software will be held to formal standards like

DO-178C that have no equivalent in the commercial world. In this part of the domain, you will learn the true meaning of high-consequence engineering, where every decision is made with the understanding that a pilot's life may one day depend on its perfection.

For an engineer transitioning from another domain, these cultural drivers can be a source of initial shock. An engineer from the pure software world might be baffled by the formal, hardware-centric rigor of configuration management. A land systems engineer, accustomed to valuing ruggedness above all, might be surprised by the fanatical, gram-by-gram war on weight. Understanding that these are not arbitrary rules, but the hard-won lessons written in the language of physics and mission failure, is the key to adapting and thriving in this unique environment. It is this mindset, combined with the core technical skills, that defines the engineers in highest demand and the major players who dominate this arena.

The Race for the Sixth Generation

The "Fifth Generation," with its focus on stealth and data fusion, is the established reality of today. But the Air Domain is defined by its relentless pursuit of the next leap forward. As a young engineer entering this world, your challenge will not be to maintain the F-35; it will be to help invent its successor. The "Sixth Generation" of air power is being designed in classified laboratories right now, and it is being defined by a revolutionary fusion of the megatrends reshaping the entire defense industry.

The future of the Air Domain will be:

Optionally Manned: The cockpit itself may become a mission module. The same airframe may be flown by a human pilot for a complex mission one day and as a fully autonomous drone the next. This requires a fundamental shift in control system architecture and a new level of trust in artificial intelligence.

A System of Systems: The future is not a single fighter, but a manned "quarterback" aircraft commanding a swarm of lower-cost, unmanned Collaborative Combat Aircraft (CCA) like the "Loyal Wingmen" we discussed in the megatrends chapter. These AI-enabled drones will fly ahead of the manned jet, acting as its sensors, its jammers, and even its weapons. The primary engineering challenge is shifting from platform design to network design by creating the resilient, secure, and high-bandwidth datalinks that allow this robotic team to function as one.

Built on a Digital Foundation: The concept of the "Digital Twin" is at the heart of the next generation of aircraft design. In a new paradigm sometimes called the "Digital Century Series," new aircraft will be designed, tested, and virtually flown thousands of times in high-fidelity simulations before any metal is ever cut. This allows for radically faster development timelines, moving from a clean-sheet design to a flying prototype in a matter of a few years, not decades. This requires engineers to be masters of their digital tools, from Model-Based Systems Engineering (MBSE) to advanced computational analysis.

Defined by the Seams: The future of the air is inextricably linked to other domains. It involves the integration of hypersonic weapons, the use of airborne directed energy systems (lasers), and a complete reliance on the Space and Cyber domains for communication, navigation, and data fusion. The most valuable aerospace engineer of the future will be the one who can operate and innovate at these critical seams.

Key Roles & Major Players

Engineers in High Demand: The Core Disciplines

The Air Domain is a vast and interdisciplinary field, but it is built on a foundation of four core engineering disciplines. While many other specialties are needed, your career will almost certainly start in one of these pillars:

Aerospace & Mechanical Engineers: These engineers are the bedrock of the Air Domain. They own the physical platform itself: the aerodynamics, the flight controls, the structures that must be both lightweight and immensely strong, the propulsion systems, and the relentless thermal challenges of high-speed flight. They are the masters of the physical forces that govern the aircraft.

Electrical & Computer Engineers: These engineers design the "nervous system" of the platform. They are the experts in RF and sensor engineering who build the powerful radars and passive detectors. They design the flight-critical digital hardware and the embedded systems that control every function of the aircraft. Their world is the physics of the electromagnetic spectrum and the logic of digital circuits.

Software Engineers: On a modern aircraft, software is the system. Computer scientists and software engineers are in massive demand to write the millions of lines of highly reliable, real-time code for everything from the flight controls to the sensor fusion algorithms. With the rise of the JADC2 "combat cloud" and AI-driven autonomy, their role is shifting from a support function to being the primary driver of capability.

The Dominant Companies: The Industrial Titans

While the ecosystem contains thousands of companies, the landscape is dominated by a handful of giants with deep histories and massive resources.

The "Big Three" Primes: Lockheed Martin (and its legendary Skunk Works® advanced development group), Northrop Grumman, and Boeing Defense are the primary system integrators. They are the ones who design and build the most advanced aircraft in the world, like the F-35 and B-21. A career here means working on massive, complex, and incredibly well-funded Programs of Record.

The Power Players (Engines): The world of advanced jet engines is dominated by two primary American players: Pratt & Whitney (part of RTX) and GE Aviation. These companies tackle the most extreme materials and thermodynamics challenges in the industry.

The Munitions Masters: Companies like Raytheon are the undisputed leaders in designing and building the "arrows," the advanced missiles and smart bombs that are as complex as the aircraft that carry them.

Government & Research: The Customers and Innovators

It's critical to remember that the ultimate customer is the U.S. government, which is also a primary hub for employment and cutting-edge research.

The Research Labs: The Air Force Research Laboratory (AFRL) is the primary driver of foundational, long-range research for the Air Force and Space Force. They partner extensively with NASA, which, while a civilian agency, often develops technologies in areas like hypersonics and advanced materials that are directly applicable to defense. For an engineer who wants to live in the world of pure R&D, these government labs are a top destination.

The Flight Test Centers: Places like Edwards Air Force Base in California are where theory meets reality. This is the world of flight test. Here, civilian government engineers work alongside military test pilots to put new aircraft through their paces, find their limits, and verify that they meet their performance requirements. It is a world of hands-on, high-stakes, and deeply practical engineering.

The Inheritance of Innovation

To be an engineer in the Air Domain is to inherit a legacy. You are stepping into a story that began with the pioneers who first broke the sound barrier and has been written, generation by generation, by the engineers who mastered the arts of supersonic flight, precision guidance, and stealth. The defining characteristic of this domain is a culture of relentless, forward-looking innovation, a deeply ingrained understanding that today's cutting-edge advantage is tomorrow's baseline.

This culture is a direct result of the unique physics of the environment and the life-or-death stakes of the mission. The fanatical obsession with weight, the strict discipline of configuration management, and the almost unimaginable rigor of designing flight-critical systems are not arbitrary rules; they are the hard-won lessons from a century of pushing machines to the absolute limits of what is possible. This is the "Air Domain Mindset," a unique blend of deep, narrow specialization and an unwavering commitment to a process where failure is not an option.

As you begin your career, you are not just taking a job; you are taking your place in that lineage. The "Fifth Generation" of aircraft, with its focus on stealth and data fusion, is the established reality. Your challenge, and the challenge of your generation of engineers, will be to invent the Sixth Generation. This will be a world defined by the seams between domains; by the fusion of artificial intelligence with flight controls, by the ability of a single pilot to command a swarm of autonomous drones, and by the creation of new materials and propulsion systems to master the brutal environment of hypersonic flight.

The Air Domain demands a unique blend of creativity and discipline, of bold vision and meticulous execution. For the engineer who is captivated by the elegant complexity of flight and who thrives on solving problems at the very edge of possibility, this arena is the place for you.

Chapter 8

The Land Domain

Mud, shock, and vibration rule the Land Domain. As the most visceral of all defense arenas, this is the world of direct, physical interaction; the domain of the soldier, the tank, and the robotic system that navigates the unforgiving and complex terrain of the real world. While an aerospace engineer battles the elegant physics of the upper atmosphere, the land systems engineer wages a brutal, close-quarters war against the laws of the ground: shock, vibration, dust, mud, extreme heat, and the ever-present, violent threat of enemy fire.

For an engineer, this domain offers a chance to work on the systems that directly protect and empower soldiers. This is a world of thick armor, powerful engines, and ruggedized hardware. It is a domain defined by a relentless focus on reliability and survivability. The systems designed here don't have the luxury of gliding through a clean, predictable environment; they must perform, without fail, in the harshest conditions on Earth. It is a field that demands a deep and intuitive understanding of mechanical forces and a profound respect for what it takes to build something that is, in a word, unbreakable.

A Legacy of Ground Innovation

The history of land warfare engineering is a story written in steel, mud, and hydraulics. It is a continuous, evolutionary struggle defined by a fundamental trinity of competing forces: firepower, mobility, and protection. For over a century, engineers have been locked in a relentless cycle, where an advance in one area (like a more powerful anti-tank weapon) immediately necessitates a breakthrough in another (like stronger armor).

The First World War introduced the world to the horror of trench warfare and the machine that would break its stalemate: the tank. These early machines were crude, unreliable behemoths, but they established the core engineering challenge that would define the next century. The Second World War was the era of mass production, where engineers at companies like Chrysler mastered the art of building thousands of reliable M4 Sherman tanks.

The Cold War turned this into a high-tech race. The challenge was to counter the massive numerical advantage of Soviet tank armies in Europe. This spurred the development of advanced composite armors (like the Chobham armor on the M1 Abrams), sophisticated fire control systems that could shoot accurately on the move, and powerful gas turbine engines that gave a 70-ton tank the agility of a much lighter vehicle.

The conflicts of the Post-Cold War era introduced a new, insidious threat: the Improvised Explosive Device (IED). This shifted the engineering focus dramatically toward crew survivability, leading to the development of V-shaped hulls that deflect blast energy and advanced suspension systems to handle the immense weight of the additional armor. This is the world that created the JLTV and the MRAP (Mine-Resistant Ambush Protected) family of vehicles.

> *Sidebar: Tactical Vehicles and Combat Vehicles are terms that are naively seen as interchangeable. Tactical Vehicles are intended to protect its operators in the event that contact is made but are meant for more logistical roles. Combat Vehicles, on the other hand, are designed to go into combat with the anticipation of making contact and taking fire.*

As an engineer entering the Land Domain today, you are stepping into this legacy. The work you do is part of the next great leap, which is defined by robotics and autonomous systems. The challenge is no longer just about building a better tank, but about building teams of manned and unmanned vehicles that can work together to dominate the future battlefield.

The Bedrock Programs: The Full Spectrum of Ground Warfare

While the iconic image of a main battle tank charging across a field is a powerful symbol of the Land Domain, it represents only the percussion section in a vast and complex orchestra. The reality of modern ground warfare is a deeply integrated symphony of capabilities, where the tank is supported by long-range artillery, empowered by a vast digital network, and enabled by a resilient logistical backbone. The work of a land systems engineer, therefore, is equally diverse. It spans the entire spectrum of this ecosystem, from the advanced materials in an individual soldier's helmet to the complex software that manages the entire battlefield.

Combat Vehicles: This is the traditional heart of the domain. It includes the M1 Abrams Main Battle Tank (General Dynamics), the versatile Stryker family of wheeled vehicles, and the highly survivable Joint Light Tactical Vehicle (JLTV) from Oshkosh Defense. The engineering is a constant balance of firepower, mobility, and protection.

Fire Support Systems: This is the world of "the king of battle": artillery. It's a domain of immense mechanical and chemical engineering challenges. It includes not only traditional cannon artillery like BAE Systems' M109 Paladin self-propelled howitzer, but also the incredibly powerful rocket and missile artillery systems like the HIMARS (High Mobility Artillery Rocket System) from Lockheed Martin, which can deliver precision effects from dozens of miles away.

Soldier & Small Arms Systems: This is engineering at its most personal and human-centric scale. It involves designing the next generation of rifles, optics, and body armor. It also includes the development of advanced soldier systems comprised of integrated headsets, radios, and heads-up displays (like the Integrated Visual Augmentation System - IVAS) that provide infantry with unprecedented situational awareness. This is a world of human factors, ergonomics, and miniaturization.

C4ISR and Network Systems: We define this as the discipline of building things. C4ISR stands for Command, Control, Communications, Computers, Intelligence, Surveillance, and Reconnaissance. It is the engineering of the ruggedized radios, satellite terminals, and tactical networks that allow a commander to see the battlefield. It includes vital systems like

Blue Force Tracking, which displays the positions of friendly units on a digital map, preventing fratricide and enabling complex maneuvers.

Logistical Systems: An army is only as good as its supply chain. This often-overlooked but critical area involves engineering the massive, heavy-duty tactical trucks (like the HEMTT family) and logistics systems required to transport fuel, ammunition, and supplies to the front line. It is a world of incredible reliability and durability challenges.

Sustainment & Modernization Programs: This is one of the largest and most stable areas of land systems engineering. The Army doesn't just buy a fleet of tanks and use them for 30 years; it is constantly upgrading, modernizing, and sustaining them. This involves massive programs to replace engines, upgrade sensors, install new communication gear, and redesign components that have become obsolete. This is a world of reverse engineering, failure analysis, and clever integration, where engineers are challenged to fit 21st-century technology into 20th-century platforms.

Core Engineering Challenges

The engineering challenges of the Land Domain are as diverse and integrated as its programs. The laws of physics are unforgiving on the ground, but they are compounded by the complexities of networking, human psychology, and the absolute necessity for rugged, failsafe reliability. A land systems engineer must be a master of their core discipline, but also a student of the many adjacent fields that contribute to a successful system. From the molecular structure of an armor plate to the architecture of a tactical cloud, the problems are deep and multi-faceted.

Survivability (The Science of Not Getting Hit and Surviving When You Do): This remains the primary concern. For a modern combat vehicle, this is a multi-layered science that goes far beyond just thick steel. It involves designing advanced, composite armors like Chobham that are layered with ceramics and polymers to defeat shaped charges. It also includes the world of Active Protection Systems (APS) which are typically miniature radar systems and shotgun-like projectiles designed to shoot down incoming anti-tank missiles moments before impact. The work involves deep materials science, but also the complex sensor fusion and split-second algorithm development needed to make an APS effective.

Mobility & Powertrains: The challenge of moving a 70-ton Main Battle Tank over punishing, unimproved terrain at 40 miles per hour is a monumental feat of mechanical engineering. The work is not just about raw power, but about reliability and endurance. As a powertrain engineer, you could be designing and integrating incredibly power-dense engines from massive gas turbines to advanced hybrid-electric drives into tightly packed engine compartments. A suspension engineer might work on advanced hydro-pneumatic suspension systems, incredibly robust components that must manage the immense static and dynamic loads of a massive vehicle while providing a stable enough platform to fire the main gun accurately while on the move.

Shock, Vibration, and Ruggedization: While other domains contend with a primary physical challenge, the Land Domain is under constant, harsh, multi-axis assault from shock and vibration. Every single component, from the thickest armor plate to the most delicate circuit board, must be designed to survive a lifetime of this punishment, such as the bone-jarring slam of a vehicle coming off a ledge, the high-frequency vibration of a track on rocky terrain, and the violent shock of a nearby blast. A mechanical engineer specializing in structural dynamics becomes an expert in vibration isolation, designing clever mounting systems with elastomeric dampers to protect sensitive electronics. An electrical engineer learns that a standard commercial laptop will fail in minutes and instead must design "ruggedized" computers with no moving parts, solid-state hard drives, and fully sealed connectors to prevent dust and water intrusion. This is the unseen, relentless enemy of every land system, and the ability to design hardware that is fundamentally unbreakable is the most respected skill in this domain.

Robotics & Autonomous Navigation: This is the new frontier of the Land Domain. The challenge is enabling robotic vehicles to navigate, perceive, and act in a chaotic, GPS-denied, and unstructured battlefield environment where the "road" might be a forest or a collapsed urban street. For a software or robotics engineer, this means working with sensor data from LIDAR, stereo cameras, and radar, and developing sophisticated SLAM (Simultaneous Localization and Mapping) and perception algorithms that can distinguish between a harmless bush, a dangerous obstacle, and a potential threat. It's an incredibly complex challenge that sits at the cutting edge of AI and machine learning research.

Network Engineering & RF Communication: This is the core of C4ISR and the "nervous system" of the ground force. A tank is blind without data, and a commander is powerless without communication. The challenge is to create a robust, mobile, ad-hoc network that can function reliably in a chaotic and actively contested electromagnetic environment. An RF engineer's work involves designing ruggedized, multi-band antennas that can survive the severe shock and vibration of a combat vehicle, while a network engineer develops clever waveforms and routing protocols that allow the network to "heal" itself, automatically finding a new path for data to flow when a node is destroyed or jammed by the enemy.

Human Factors & Ergonomics: Unlike a satellite, every system in this domain has a human at its core. As a human factors engineer, your job is to be the advocate for that soldier. Your work might involve building a physical mockup of a new vehicle's interior to ensure a soldier wearing 60 pounds of gear can get in and out quickly or designing the user interface for a new heads-up display to present critical data without overwhelming the user in the heat of combat. It is a deep, interdisciplinary challenge that blends mechanical engineering, software design, industrial design, and even human psychology to ensure that the system is not just powerful, but usable and intuitive under the most extreme stress imaginable.

A Day in the Life: Three Perspectives on the Ground

The daily experience of a Land Domain engineer is deeply rooted in the physical world, and the problems they solve can span the entire lifecycle of a platform, from a clean-sheet design to a 40-year-old vehicle that needs a new lease on life.

The Design Engineer:

A mechanical engineer like "Alex" might spend their week using a specialized FEA tool like LS-DYNA to run a highly complex, non-linear blast analysis on a new hull design. They simulate an explosion down to the microsecond, meticulously analyzing how the shockwave propagates and how the steel structure deforms, all to ensure the crew inside would survive. Their world is a quiet, intense dialogue with the violent physics of explosions, a cycle of simulation, analysis, and refinement. They are trying to create the most survivable vehicle possible within the constraints of weight and cost, and their primary collaborators are other designers and analysts.

The Robotics Engineer:

A software or electrical engineer might spend their week in the field, riding in the back of a prototype robotic vehicle with a laptop open. As the vehicle navigates a difficult off-road course using its sensors, the engineer is tuning the perception and pathfinding algorithms in real-time. Their lab is the mud and dust of the proving ground, and their work is a constant, rapid cycle of coding, testing, and immediate feedback from the real world. They are living on the cutting edge, trying to teach a machine how to think and see in a chaotic environment, and they work hand-in-glove with the vehicle operators and test technicians.

The Sustainment Engineer:

This is one of the most common and critical roles in the entire industry. Let's imagine "Sarah," an electrical engineer tasked with a seemingly impossible problem: a critical computer on a 25-year-old vehicle is failing, and the original manufacturer of a key microchip went out of business a decade ago. Sarah's week is a masterclass in engineering forensics. She starts by digging through old, scanned PDF schematics to understand the original circuit design. She then works with component engineers to find a modern, available chip that *might* work. The new chip uses a different voltage and has a different physical footprint, so Sarah spends her "deep work" time designing a small adapter circuit board to make them compatible. Her collaborative work involves negotiating with the manufacturing floor to get her small, high-priority board built, and working with test technicians to prove that her solution works without affecting the other systems on the vehicle. This is the gritty, essential work of keeping the fleet running.

What Makes The Land Domain Unique?

If the Air Domain is ruled by the tyrant of gravity, the Land Domain is governed by a harsh gang of thugs: shock, vibration, dust, and mud. Every design decision, every cultural norm, and every engineering trade-off is a response to this relentless, physical assault. This fosters a unique engineering culture and mindset that stands in stark contrast to other domains.

Reliability is the Highest Virtue: While an aerospace engineer obsesses over weight, a land systems engineer obsesses over reliability. The systems designed for this domain must be operated and, crucially, maintained in the most austere conditions imaginable like freezing mud or a baking desert by a 19-year-old soldier with a wrench. This fosters a deep cultural respect for designs that are robust, simple, and maintainable. A complex, high-performance system that cannot be fixed in the field with basic tools is a liability, not an asset. You will quickly learn that the design philosophy is often "simpler is better."

"Hardware Is Hard" Mentality: This is a domain of steel, hydraulics, and high-power engines. While software and electronics are a critical and growing component, there is a deep, cultural respect for the physical, "heavy engineering" of the platform itself. A successful engineer in this space must be comfortable with the tangible world. In your design reviews, you will be expected to not only defend your FEA model but also to have an intelligent conversation with the manufacturing engineers about weld callouts, fixturing, and fabrication realities. Your reputation will be built not just on your analysis, but on your ability to walk a factory floor and understand how your digital model will be turned into thousands of pounds of steel.

The Primacy of Testing and the Proving Ground: While all domains rely on simulation, the Land Domain has a deep-seated cultural belief in the ultimate authority of physical testing. The environment is simply too complex and chaotic to be perfectly modeled. Therefore, the ultimate arbiter of a design's success is not the simulation, but its performance at a test facility like the Aberdeen Proving Ground or the Yuma Proving Ground. You will find that requirements are often verified through grueling, months-long Durability and Reliability testing, where vehicles are driven for thousands of miles over punishing courses designed to find and exploit their breaking points.

A Focus on the Human Element: Unlike an autonomous satellite or a fire-and-forget missile, the systems in this domain are operated, maintained, and often *lived in* by soldiers. This creates a profound focus on human factors engineering. In your design process, you will be constantly challenged with questions that don't exist in other domains: Can a soldier wearing bulky gear get in and out of the vehicle? Are the controls intuitive enough to be used under the extreme stress of combat? A great land systems engineer never forgets that their ultimate end-user is a person whose life depends on their work, and they design accordingly.

The Power of the "-ilities": More than in any other domain, the so-called "-ilities" are king here. Manufacturability, Maintainability, Sustainability, and Affordability are often just as important as the raw performance specifications. A brilliant design that is too expensive to build in large numbers or too difficult to repair in the field is a failed design. This is a domain of pragmatic trade-offs.

An engineer transitioning from the Air Domain might be surprised by the relative lack of focus on weight and the intense focus on cost and maintainability. An engineer from the pure software world might be challenged by the deep respect for hardware and the realities of the manufacturing floor. Understanding these cultural drivers is the key to succeeding in this rugged and rewarding field.

Key Roles & Major Players

The land systems industrial base is a gritty, hands-on world of heavy manufacturing and, increasingly, cutting-edge software and robotics. Understanding the key players and the engineers they are desperately seeking is critical to finding your place on the modern battlefield.

Engineers in High Demand

The Land Domain demands a unique blend of the traditional and the futuristic. While mechanical engineering remains the bedrock, the push toward autonomy is creating massive demand for new skill sets.

Mechanical Engineers: This discipline remains the heart of the Land Domain. They are the masters of "heavy engineering": the design of massive, welded steel structures, powerful powertrains and transmissions, and advanced suspension systems. Their world is defined by survivability, reliability, and manufacturability.

Electrical & Robotics Engineers: As vehicles become more electrified and autonomous, the demand for this talent is exploding. Electrical engineers work on everything from robust power distribution and hybrid-electric drive systems to the design of ruggedized sensor and communication payloads. Robotics engineers are the masters of mechatronics, integrating the motors, sensors, and actuators that bring an unmanned system to life.

Software Engineers: This is the fastest-growing discipline in the Land Domain. Computer scientists and software engineers are needed to write the AI/ML algorithms for autonomous navigation and target recognition, the networking protocols for manned-unmanned teaming, and the complex software for the C4ISR systems that form the battlefield's nervous system.

The Dominant Companies: The Industrial Titans

While thousands of smaller companies supply components, the world of primary combat vehicle design is dominated by a few established titans.

The Combat Vehicle Giants: The two undisputed leaders in the design and production of tracked and wheeled combat vehicles in the United States are General Dynamics Land Systems (maker of the M1 Abrams tank and the Stryker vehicle) and BAE Systems (maker of the M109 Paladin artillery system and the Armored Multi-Purpose Vehicle).

The Tactical Wheeled Vehicle Leader: The world of heavy and medium tactical trucks, the logistical backbone of the Army, is led by companies like Oshkosh Defense, the maker of the highly survivable JLTV (Joint Light Tactical Vehicle) and M-ATV (Mine Resistant Ambush Protected All-Terrain Vehicle).

Government & Research

The U.S. Army is the ultimate customer for all these systems, and it also serves as the central hub for the research and testing that drives the entire industry forward.

The R&D and Engineering Centers: The primary driver of R&D in this space is the U.S. Army's Combat Capabilities Development Command (DEVCOM). Within DEVCOM, the Ground Vehicle Systems Center (GVSC), located in the Detroit, Michigan area, is the heart of government research and engineering for combat vehicles. This is where Army civilian engineers work on the next generation of technologies that will eventually make their way into future platforms.

The Proving Grounds: A design's success is not determined in a simulation, but on the brutal, punishing courses of the Army's test facilities. Places like the Aberdeen Proving Ground in Maryland and the Yuma Proving Ground in Arizona are where vehicles are driven for thousands of miles over punishing terrain to find their breaking points. A career here as a civilian test engineer is a hands-on, hardware-rich role focused on verifying the reliability and performance of systems in the real world.

Engineering for the Ground Truth

To be an engineer in the Land Domain is to be intimately connected to the "ground truth" of national defense. It is a world without the glamour of supersonic flight or the mystery of space, but it is a world of profound and tangible purpose. The systems you design are the armored shells that protect soldiers, the robotic partners that scout ahead into danger, and the reliable vehicles that ensure the mission can be accomplished.

This culture is a direct result of the unique physics of the environment. The deep respect for reliability, the "hardware is hard" mentality, and the intense focus on maintainability and the human element are not arbitrary rules; they are the hard-won lessons from over a century of designing machines that must function in the most severe conditions imaginable. This is the "Land Domain Mindset," a unique blend of pragmatism, a deep respect for the physical world, and a relentless focus on building things that simply do not break.

As you begin your career, you are not just taking a job; you are becoming a steward of the soldier's primary tool. The era of the standalone combat vehicle is ending. Your challenge will be to invent the future of ground warfare. This means pioneering the world of manned-unmanned teaming, creating robotic wingmen for armored vehicles that can scout ahead into danger. It is a world defined by the complicated seam between ruggedized hardware and intelligent software, by networks that can survive the chaos of close combat, and by the ethical challenge of designing autonomous combat systems that can be trusted as true partners for soldiers on the battlefield.

The Land Domain demands a unique breed of engineer: a pragmatist who respects the unforgiving physics of the earth, who can blend advanced analysis with a hands-on feel for hardware. It is for the engineer who finds profound satisfaction in building things that are rugged, powerful, and unquestionably tough. In a world of elegant theories, this is the domain of hard-won ground truth, and there is no more direct way to apply your skills to the safety and effectiveness of the men and women on the front line.

Chapter 9

The Sea Domain

Welcome to the ultimate systems integration challenge. In the Sea Domain, the naval engineer must design a small, self-sufficient city. It must generate its own electrical power grid for thousands of residents, produce tens of thousands of gallons of fresh water each day from the ocean, manage its own waste, and withstand the harshest environmental conditions on the planet, from the crushing ice of the Arctic to the 50-foot waves of a typhoon in the Pacific. Now, imagine that this city must also be a mobile military base, a functioning airport, a nuclear power plant, and a high-tech command center, all while moving at 30 knots through a relentlessly corrosive saltwater environment that is actively trying to dissolve it.

Welcome to the world of naval engineering. Aircraft carriers, destroyers, and submarines, the primary platforms here are some of the most complex machines ever created by humankind. An engineer in this domain is not just designing a single component; they are weaving that component into an intricate, three-dimensional landscape of a thousand other systems. Every decision is a trade-off between space, weight, power, and cooling, all within the strict, unyielding confines of a steel hull where a single misplaced pipe can cause a months-long delay. Such a domain calls to those who see the big picture, who thrive on complexity, and who understand that the whole is always greater, and more challenging, than the sum of its parts.

A Legacy of Naval Innovation

The history of naval engineering is a story of scale, power, and complexity. For centuries, the core challenge was simple: build a bigger ship with bigger guns. This culminated in the mighty battleships of the early 20th century, like the HMS Dreadnought, which were floating steel fortresses that defined national power. But the attack on Pearl Harbor proved in a single morning that a new king of the seas had arrived: the aircraft carrier. The ability to project air power from a mobile, sovereign base across the ocean became the new measure of a superpower, shifting the primary engineering challenge from gunnery to aviation integration.

The Cold War was the golden age of naval engineering innovation, driven by the existential need to counter a massive and growing Soviet submarine threat. This spurred two parallel technological revolutions that define the modern Navy. First was the perfection of the nuclear-powered submarine. Championed by the legendary and notoriously demanding Admiral Hyman G. Rickover, this created a true submersible for the first time in history. No longer dependent on diesel engines that needed air, these predators could stay hidden in the depths for months at a time, limited only by the endurance of the crew. The engineering challenges of building a safe, silent, and reliable nuclear reactor that could fit inside a pressure hull were immense and foundational to the Navy's global reach.

Second was the birth of the guided missile and the Aegis Combat System. As anti-ship missiles became faster and smarter, the Navy realized that traditional guns were no longer sufficient for fleet defense. The solution was a revolutionary, integrated system that linked powerful new phased-array radars (the SPY-1), high-speed computers, and a new generation of surface-to-air missiles into a single, cohesive "shield." This turned the ship itself into a single, integrated weapon system and marked the beginning of the digital age for naval warfare.

As an engineer entering the Sea Domain today, you are stepping into this legacy. The work you do is part of the next great leap, which is defined by autonomy and networking. The challenge is no longer just about building a bigger, better ship, but about creating a networked fleet of manned and unmanned vessels that can work together, sharing sensor data to control vast areas of the ocean, from the surface to the seabed.

The Crown Jewel Programs

The work in this domain is focused on platforms of incredible scale and complexity, the cornerstones of the U.S. Navy.

The Power Projectors

These are the primary assets that project American power across the globe, capable of operating for months at a time, thousands of miles from home.

The Ford-class Aircraft Carrier: A 100,000-ton marvel of engineering, a true "supercarrier" that serves as a mobile piece of sovereign U.S. territory. The defining feature of this new class, built by HII's (Huntington Ingalls Industries) Newport News Shipbuilding, is its leap into the electrical age. It replaces the classic steam-powered catapults with the powerful and precise Electromagnetic Aircraft Launch System (EMALS). This single change required a complete redesign of the ship's power generation and distribution systems, a monumental engineering feat involving massive flywheel generators and advanced power electronics capable of handling incredible energy surges.

The Virginia-class Submarine: A predator designed for the deep. These nuclear-powered attack submarines, built by General Dynamics Electric Boat, are paragons of stealth and endurance. The engineering challenge is immense: designing a life support system that can generate oxygen from seawater and scrub the air for a crew of over 100 in total isolation, creating a nuclear reactor that is so quiet it is acoustically undetectable, and building a hull that can withstand the crushing, bone-breaking pressure of the deep ocean.

The Shields of the Fleet

The primary mission of the surface fleet is to defend the high-value assets and control vast areas of the ocean. This mission is accomplished not by a single ship, but by a networked shield.

The Arleigh Burke-class Destroyer & The Aegis Combat System: Serving as the workhorse of the U.S. Navy, built by both General Dynamics Bath Iron Works and HII (Huntington Ingalls Industries). While the ship itself represents a masterpiece of naval architecture, it derives its lethal power from its heart: the Aegis Combat System. Developed by Lockheed Martin and Raytheon, Aegis is the revolutionary "brain and shield" of the fleet. Rather than a single piece of hardware, Aegis functions as a deeply integrated network of powerful phased-array radars (like the SPY-6), high-speed computers, and a versatile arsenal of guided weapons. As a single, integrated weapon system, the Aegis destroyer can track hundreds of targets simultaneously, from sea-skimming cruise missiles to ballistic missiles outside the atmosphere and engage them. For software and electrical engineers, working on the Aegis Combat System is working on the digital fortress that protects the entire fleet.

The Unmanned Vanguard

Just as in the Air and Land domains, the future of the Sea Domain is autonomous. The Navy is heavily investing in a new generation of unmanned platforms designed to extend the reach and sensing capability of the manned fleet at lower cost and with less risk to sailors.

Unmanned Undersea Vehicles (UUVs): These are autonomous, robotic submarines, like Boeing's Orca XLUUV (Extra-Large UUV). The engineering challenge is immense: designing a vehicle that can navigate autonomously for weeks or even months in the dark, unmapped deep ocean, manage its own power systems, and perform missions like intelligence gathering or mine-hunting without any human intervention.

Unmanned Surface Vessels (USVs): From small, sensor-laden boats to massive, destroyer-sized platforms like the Overlord program, these are the robotic scouts of the surface fleet. The engineering challenge involves not only autonomous navigation that can safely handle complex sea traffic and weather, but also creating the resilient, long-range communication links that allow them to be effective members of the networked fleet.

Core Engineering Challenges

Systems Integration: We define this as the discipline of building things of the naval engineer. The inside of a modern destroyer is one of the most densely packed, complex industrial environments ever created. As an engineer, your job is to solve a high-stakes, three-dimensional jigsaw puzzle. Using a specialized CAD tool like ShipConstructor or CATIA, you might be a marine engineer tasked with routing massive cooling water pipes, high-voltage electrical cableways, and vital firefighting systems through the same small compartment, all while leaving enough room for a 20-year-old sailor to access a critical valve in a rolling sea. Every decision is a negotiation; moving a pipe one inch to the left might solve your problem but might then interfere with an electronics rack, creating a cascading effect on a dozen other systems. This is the art of the possible in a world of unyielding steel bulkheads.

Hydrodynamics & Structures: The sheer scale of a 9,000-ton destroyer or a 100,000-ton aircraft carrier presents a monumental structural challenge. For a surface ship, a naval architect's work involves analyzing the global hull girder strength using massive FEA models to ensure the ship can withstand the immense, millions-of-cycles stress of waves over a 40-year life. For a submarine, the problem is even more extreme. The structural engineer must design a pressure hull from high-strength steel that can withstand the equivalent of a ton of pressure on every square inch of its surface, where a single microscopic flaw in a weld, invisible to the naked eye, could lead to catastrophic, instantaneous failure in the deep ocean.

Corrosion & Materials Science: From the moment a ship touches the water, the ocean is actively trying to dissolve it. A naval engineer's career is a constant battle against corrosion. This is a multi-layered defense fought with deep materials science expertise. It involves specifying advanced marine coatings and paints, carefully selecting exotic alloys like copper-nickel (cupronickel) for seawater piping systems due to their inherent corrosion resistance, and designing and maintaining complex cathodic protection systems, which use carefully controlled electrical currents to trick the saltwater into eating a sacrificial piece of metal instead of the steel hull itself.

Acoustic Stealth: For a submarine, silence is not just a feature; it is its primary form of survival. An immense amount of engineering effort, spanning nearly every discipline, is dedicated to making them acoustically invisible. A mechanical engineer might work on designing the massive, multi-ton vibration-damping "rafts" on which a submarine's powerful engines and generators are mounted, acoustically isolating their vibrations from the hull. An electrical engineer works to design power converters that have no audible "hum." A materials engineer designs the specialized anechoic tiles that cover the entire hull, which are designed to absorb the sound waves of an enemy's active sonar, preventing an echo from returning. Their lab is a quiet room full of sensitive hydrophones and spectrum analyzers, and their enemy is a single, unwanted decibel of noise.

A Day in the Life: Three Perspectives

The daily life of a naval engineer is a masterclass in managing complexity, whether digital or physical.

The Marine Engineer:

A mechanical engineer like "Alex" might spend their week in a 3D CAD tool like ShipConstructor or CATIA, meticulously planning the layout of a ship's auxiliary machine room. Their work is a high-stakes game of three-dimensional chess. They are routing massive pipes for cooling water, fuel, and steam around structural beams, high-voltage cableways, and ammunition magazines. Every decision is a negotiation. Moving a pipe one inch to the left might solve their problem but create a new one for the electrical team. Their day is a mix of focused design work and collaborative meetings, ensuring that every component is not only in the right place, but is also accessible for a 20-year-old sailor to perform maintenance in a rolling sea.

The Combat Systems Engineer:

An electrical or software engineer could be in a shore-based integration facility, a massive lab where the real hardware and software of the Aegis system are tested. Their job is to troubleshoot a complex integration issue between a new radar software update and the missile launching system. This is not a simple debugging task. It involves deep collaboration between multiple companies and engineering teams to trace a single, elusive bug through millions of lines of code and across a dozen different electronic systems. Their work is a patient, forensic investigation to ensure the ship's "nervous system" is flawless.

The Naval Architect:

A structural engineer or naval architect might be using a specialized FEA tool to analyze the structural response of an aircraft carrier's landing deck to the violent, repeated impact of a 30-ton aircraft. But their analysis goes deeper. They must also consider how that immense, localized stress affects the entire ship's structure, its stability, and its long-term fatigue life. Their work is the ultimate "big picture" analysis, ensuring the fundamental integrity and safety of the floating city itself.

What Makes The Sea Domain Unique?

If gravity rules the air and the physical earth rules the land, the Sea Domain is governed by two ruthless masters: the unforgiving ocean itself and the unyielding tyranny of the steel box. The sheer scale of its platforms and the immense power of their environment fosters a unique engineering culture distinct from all others.

Integration is Everything: More than any other domain, naval engineering is the science of the whole. A ship is arguably the most complex single machine ever created, and every cubic foot is a battleground between competing systems. You are never just a "propulsion engineer"; you are a ship engineer who specializes in propulsion. You will be forced to develop a deep appreciation for how your system draws power, how much cooling it needs, how it is physically mounted, and how it impacts the ship's overall stability. This holistic, "ship-level" thinking is the most prized skill. Your work will be a constant, collaborative negotiation, because a one-inch change to your component could create a show-stopping interference with a high-voltage cableway, a structural beam, or another engineer's system.

A Culture of Ownership and Longevity: Ships and submarines are designed to be in service for 30, 40, or even 50 years. This creates a unique "cradle-to-grave" engineering culture that is rare in the modern world. It is not uncommon for a senior engineer to have spent their entire career working on a single class of ship, from its initial concept design, through the sea trials of the first vessel, and into its decades-long sustainment and modernization. This fosters a profound, generational sense of ownership that is passed down from mentors to new hires. When you design a component, you are doing so with the understanding that the company, and perhaps even you, will be responsible for it half a century from now.

The Tyranny of Space, Weight, and Stability: While an aerospace engineer obsesses over weight, a naval engineer must wage a three-front war. Every kilogram and every cubic foot is meticulously tracked. But naval engineers have a third, critical constraint that pilots don't worry about: stability. Every piece of equipment added high up on a ship raises its center of gravity, making it less stable in heavy seas. This is not a theoretical concern; it is a fundamental issue of safety and seaworthiness. You will discover that a request to add a new 500-pound antenna topside will trigger a detailed naval architecture review, analyzing its impact not just on power and cooling, but on the ship's fundamental righting moment in a storm.

Designing for the Maintainer at Sea: The systems you design will not be repaired in a clean, well-lit depot. They will be maintained by a 20-year-old sailor, often thousands of miles from the nearest port, in the middle of a storm with a limited set of tools. This reality places an incredible and non-negotiable emphasis on maintainability and reliability. A brilliant, high-performance system that is impossible to repair at sea is a failed design. In every design review, you will be asked the fundamental question: "How will a sailor fix this?" This forces a culture of pragmatic, robust design, where accessibility for maintenance and the simplicity of a repair procedure are just as important as the system's raw performance.

Key Roles & Major Players

The naval industrial base is a unique and specialized world, a handful of massive, legacy institutions that build the most complex machines on Earth. Understanding who they are and what kind of engineers they hire is the first step in charting a course for a career in this domain.

Engineers in High Demand

Building a floating city requires a vast array of talent, but the work is centered around a few key disciplines who must work in constant collaboration.

Mechanical & Marine Engineers: These engineers are the masters of the "3D puzzle." They design the massive propulsion systems (whether nuclear or conventional), the powerful HVAC systems, and the intricate networks of piping that carry everything from jet fuel to fresh water throughout the ship. They are the ultimate systems integrators of the physical world.

Electrical Engineers: A modern warship is an electrical power plant. EEs design the ship's entire electrical grid, from the massive generators to the complex power distribution and conversion systems that provide the clean, stable power needed for sensitive radars and combat systems. Their challenge is managing megawatts of power in a dense, vibrating, and corrosive environment.

Naval Architects & Structural Engineers: These are the specialized masters of the field. They are the owners of the ship's fundamental design: its hydrodynamic shape, its overall structural integrity, and its stability. Their work is the ultimate "big picture" analysis, ensuring the floating city itself is safe, sound, and survivable.

Nuclear Engineers: For the world of aircraft carriers and submarines, "Nucs" are an essential and elite group. Often trained through highly specialized Navy-affiliated programs, they are responsible for the design, operation, and safety of the ship's nuclear reactors.

The Dominant Companies

The world of naval construction is dominated by a few key players who form the backbone of the industrial web.

The Shipbuilders: The business of building the massive hulls of aircraft carriers, submarines, and destroyers is a world of giants. The two largest are HII (Huntington Ingalls Industries), which includes Newport News Shipbuilding in Virginia (the sole builder of U.S. aircraft carriers), and Ingalls Shipbuilding in Mississippi, and General Dynamics, which owns both Electric Boat in Connecticut (a primary builder of submarines) and Bath Iron Works in Maine (a primary builder of destroyers).

The Combat System Integrators: The "brains and shield" of the fleet are designed by the primary defense electronics houses. Raytheon and Lockheed Martin are the undisputed leaders here, developing the powerful radars, sonar systems, and the complex software of the Aegis Combat System.

Government & Research

The U.S. Navy is not just the customer; its own network of engineering and research centers is the intellectual heart of the entire enterprise.

The Systems Commands: Organizations like Naval Sea Systems Command (NAVSEA) are the primary government entity that oversees the design, construction, and maintenance of the entire fleet. Working here as a civilian engineer means you are the "smart buyer," the government's technical expert who works with both the fleet and the contractors to define requirements and ensure the final product meets the Navy's needs.

The Warfare Centers: The Navy's own R&D labs are powerhouses of innovation. Places like the Naval Undersea Warfare Center (NUWC) based in Newport, RI and Keyport, WA are the world's leading experts in submarine acoustics and torpedo technology. The Naval Surface Warfare Center (NSWC) is composed of eight locations, each with its own focus. NSWC Carderock, for instance, focuses on ship design and engineering and is home to one of the world's largest ship model testing basins. For an engineer who wants to be close to the core mission and the foundational science, these government labs are a premier career destination.

Engineering a Floating Fortress

To be an engineer in the Sea Domain is to be a master of complexity. It is a world where you contribute to building one of the most sophisticated, powerful, and survivable systems ever created. The platforms you help design are not just vehicles; they are sovereign pieces of national territory, mobile embassies, and the primary instrument of global power projection. They are floating fortresses that must function, without fail, for half a century.

This culture is a direct result of the immense challenge of operating on and under the world's oceans. The deep focus on systems integration, the long-term ownership of a platform, and the absolute necessity of designing for reliability and maintainability are the hard-won lessons of centuries of naval warfare. This is the "Sea Domain Mindset."

As you begin your career, you are stepping into a legacy of immense achievement, a story that began with wooden ships and iron men.

The era of the standalone warship is evolving into the era of the networked, multi-domain fleet. Your challenge will be to invent the future: to integrate autonomous unmanned vessels that sail alongside manned ships, to design the next generation of silent, undetectable submarines, and to develop the combat systems that can defend against hypersonic threats. The engineer who thrives on immense complexity and wants to contribute to a platform of unparalleled scale and significance will thrive in this domain.

Chapter 10

The Space Domain

In the Space Domain, you are governed by the absolute. You face an environment of perfect, hard vacuum that will boil away lubricants and weld metals together. It is a realm of savage temperature swings, from the cryogenic cold of shadow to the searing heat of direct sunlight, and it is saturated with a constant, invisible rain of high-energy radiation that is lethal to conventional electronics. The fundamental challenge of engineering for this domain can be summarized in a single, brutal phrase: The Tyranny of the Rocket Equation. Every single pound of mass you want to launch into orbit requires a proportional increase in fuel to get it there, and every pound costs thousands of dollars in energy, rocketry, and risk. This unforgiving law of physics dictates that every ounce of payload requires an exponential increase in fuel, making mass the ultimate enemy.

Welcome to the world of space engineering. It is a domain of extreme, almost fanatical reliability, where you often have only one chance to get it right. There are no repair shops in orbit. The systems you design must operate flawlessly, often for decades, making them some of the most reliable machines ever built by humankind. For an engineer, this is a field of unparalleled precision and elegance. It is the art of creating incredibly complex, lightweight, and dependable systems that serve as the eyes, ears, and voice of the nation from the ultimate high ground.

A Legacy of Orbital Innovation

The history of space engineering is a story of breathtaking speed and audacious ambition, born from a moment of profound national shock and fueled by the existential pressures of the Cold War. To understand the modern Space Domain, you must first understand the legends of its creation.

The Thunderclap of Sputnik

The story begins in the pre-dawn hours of October 4, 1957. While America slept, the Soviet Union launched a rocket from the Baikonur Cosmodrome in Kazakhstan. Its payload was an 184-pound polished metal sphere named Sputnik 1. As it circled the globe, it emitted a simple, persistent radio beep. That beep, as tiny as it was, was a thunderclap that echoed in every corner of the United States. It was broadcast on radios worldwide, a constant, orbiting reminder that the Soviet Union had achieved a stunning technological first. For the American public and its leaders, it was a terrifying announcement that they were no longer protected by two vast oceans; a potential adversary could now overfly the United States with impunity, and if they could launch a satellite, they could surely launch a nuclear warhead.

The "Sputnik crisis" was a profound technological and psychological blow, instantly kicking off the first Space Race. This Cold War competition, fueled by the existential fear of Soviet dominance in space, became the primary driver of innovation for decades. In response, the U.S. government acted with incredible speed. In 1958, two new organizations were created that would define the American space enterprise for the next half-century. The first was NASA (National Aeronautics and Space Administration), a civilian agency formed from the old NACA (National Advisory Committee for Aeronautics). Its mission was to lead the public-facing scientific and human exploration efforts, to win the "hearts and minds" of the world in a peaceful competition. The second, created in deep secrecy, was the National Reconnaissance Office (NRO), a joint CIA-Air Force organization tasked with a single, urgent mission: to build and operate the nation's spy satellites to find out exactly what the Soviets were doing.

GPS: The Side Effect of Sputnik

Interestingly enough, Sputnik led to more than just the space race; it was the catalyst for GPS. Scientists at the Johns Hopkins Applied Physics Laboratory (APL) found that by using the Doppler Effect, they could track the Soviet satellite. This was accomplished because the radio waves varied with the satellite's velocity and path relative to the receiver. As this tracking system was improved, the question was posed: If the satellite's position is known, could the location of the receiver be identified? This question led to early iterations of what would later become the Global Positioning System (GPS), the Transit System, developed by APL and ARPA (later DARPA).

There is an interesting theme throughout this field where technologies are found to have dual uses, leading to even more profound breakthroughs. Another that comes to mind that is too interesting to leave out is the inspiration for Hubble. I once attended a guest lecture at Johns Hopkins University where we were given the opportunity to hear the story of the KH-11 Key Hole satellite, a CIA and NRO spy satellite first launched in 1976. During the operation of KH-11, someone had asked what if instead of looking down at the ground, we looked to the stars? This gave birth to the concept that would become the Hubble Space Telescope, and according to many experts, the designs are extremely similar besides some modernization, removal of classified capabilities, and optimization for its use case.

As a thought experiment for the reader, look at how devices you come across and ask if that was their original use. And to further expand on that experiment, think of novel applications for existing technologies.

The Race to the Moon and the Eyes in the Sky

The decade that followed was a stunning cascade of engineering achievements, driven by the clear, audacious goal set by President John F. Kennedy in his famous 1961 speech: to commit the nation to "landing a man on the Moon and returning him safely to the Earth" before the decade was out. This was not a vague aspiration; it was a deadline that forced the invention of entirely new technologies and management techniques. The Apollo program was a monumental feat of systems engineering. Under the leadership of figures like Wernher von Braun at NASA's Marshall Space Flight Center, the massive Saturn V rocket was born. At the time of publishing, it remains the most powerful rocket ever successfully flown, a 36-

story behemoth that could launch 130 tons into orbit. At the same time, at the MIT Instrumentation Laboratory, engineers like Margaret Hamilton pioneered the very concept of software engineering to create the Apollo Guidance Computer. It was one of the very first to use integrated circuits, and the immense demand from this program and the parallel Minuteman ICBM program effectively created the modern microchip industry in what would become Silicon Valley.

Simultaneously, in the deep black world of the NRO, another, equally impressive revolution was taking place. The engineers of the CORONA program were solving the problem of reconnaissance with breathtaking mechanical ingenuity. These satellites, built by Lockheed and ITEK, took high-resolution pictures of the Soviet Union on physical film. The satellite would then eject the exposed film canisters in a small reentry vehicle, a "film bucket." This bucket would plummet back to Earth, deploy a parachute, and be snagged in mid-air over the Pacific Ocean by a C-130 aircraft trailing a grappling hook. It was a stunningly complex and successful system that, for the first time, gave U.S. leaders a clear and accurate picture of their adversary's military capabilities, a fact that arguably prevented the Cold War from turning hot.

The New Space Race

For many years after the Apollo program and the end of the Cold War, space became a relatively stable and uncontested domain. It was the exclusive playground of a few superpowers. That reality has now been fundamentally and irrevocably changed. We are now in a new, more complex Space Race. This race is not just against one adversary, but is a multi-faceted competition driven by new threats and new opportunities. The rise of near-peer adversaries with advanced anti-satellite (ASAT) capabilities, from ground-based missiles to co-orbital jammers and directed energy weapons, has transformed orbit into a contested environment. At the same time, the explosive growth of the commercial space industry, led by pioneers like SpaceX, has shattered the old launch monopolies and driven down the cost of access to space. The challenge for the modern defense engineer is no longer just getting to space; it is about operating, defending, and maneuvering in a domain that is more crowded, more competitive, and more dangerous than ever before.

The Crown Jewel Programs

The work in the Space Domain is focused on creating and maintaining the constellations of satellites that form the invisible backbone of modern life and national security. These programs are best understood not by platform, but by the critical, unwavering mission they provide to the nation.

Positioning, Navigation, and Timing (PNT)

The Global Positioning System (GPS): Perhaps the most impactful engineering program in modern history. Owned by the U.S. government and operated by the U.S. Space Force, this constellation of dozens of satellites provides the precise timing and navigation signals that enable everything from your car's navigation system to global financial transactions and the guidance of precision munitions. The core engineering challenge is maintaining a network of orbiting atomic clocks so stable and so accurate that their timing signals must be constantly corrected for the effects of both special and general relativity, as predicted by Einstein, lest the entire system drift into uselessness.

Overhead Persistent Infrared (OPIR)

Next-Generation OPIR: These are the silent, unblinking eyes that watch for global threats. A constellation of highly classified satellites in geosynchronous orbit, their primary mission is to detect the intense heat signature of a missile launch anywhere on the globe, providing precious minutes of early warning to national leaders. The engineering here is at the absolute cutting edge of infrared sensor technology, requiring the design of massive, exquisitely sensitive space telescopes that must be cryogenically cooled to operate, all while surviving for decades in the harsh radiation environment of high orbit. These systems are critical assets for both the Space Force and the broader Intelligence Community.

National Reconnaissance

National Reconnaissance Satellites: Operating in deep secrecy, this vast and highly sophisticated portfolio of assets is designed, built, and flown by the National Reconnaissance Office (NRO). These are the nation's "spy satellites," providing the critical intelligence that informs the President and other national leaders. The engineering challenges are on the absolute frontier of technology. This is the world of massive, space-based telescopes with mirror optics polished to a perfection that rivals the Hubble, and of enormous, deployable mesh antennas that can pick up the faintest radio signals from the surface of the Earth.

Core Engineering Challenges

The engineering challenges of the Space Domain are unlike those in any terrestrial environment. They are a direct consequence of the immutable laws of physics that govern orbit and the brutal nature of space itself. An engineer in this field must become a master of managing extreme forces, extreme temperatures, and extreme distances, all while fighting a relentless war against mass. These core challenges are the fundamental problems that have driven innovation in the field for over sixty years, and they are the technical puzzles you will be tasked with solving.

Astrodynamics & Propulsion: Astrodynamics constitutes the foundational science of getting into orbit and maneuvering once you're there. It's not just the brute-force rocket science of launch, but also the delicate finesse of high-efficiency, low-thrust electric propulsion (like Hall-effect thrusters) that "sip" xenon fuel to make tiny, precise orbital adjustments, extending a satellite's life by years or even decades.

Extreme Thermal Management: A satellite is in a constant, violent thermal battle. The side facing the sun can be boiling at over 120°C (250°F), while the side facing the cold of deep space is freezing at -150°C (-240°F). The thermal engineer's job is a mastery of thermodynamics without convection, designing complex systems of heat pipes, radiators, and the iconic multi-layer insulation (the "gold foil") to keep sensitive electronics within their narrow, life-sustaining operating temperature range.

Radiation Hardening ("Rad-Hard"): Once you hit the Van Allen Belt in Medium Earth Orbit (MEO), space is filled with a constant storm of high-energy particles that are lethal to standard commercial electronics. A massive engineering effort goes into designing and fabricating specialized, expensive "rad-hard" circuits. This involves using heavier shielding, redundant logic, and exotic semiconductor materials to withstand this constant bombardment without failing, and it is a primary driver of the high cost and long development times of space hardware. Much of this work requires a deep understanding of the environmental considerations to balance protection with mass savings. For example, satellites in Geosynchronous Equatorial Orbit (GEO) and Low Earth Orbit (LEO) orbits are much more protected from radiation than those in High-Earth Orbit (HEO). Those in HEO experience less radiation than GPS satellites which live in a half-GEO orbit within MEO, constantly within the severe conditions that the Van Allen Belt presents.

Structures & Lightweight Materials: The relentless war against mass, driven by the tyranny of the rocket equation, forces incredible innovation. Satellite structures are masterpieces of optimization, often using exotic composite materials, meticulously designed honeycomb panels, and additively manufactured (3D printed) titanium brackets to create designs that are feather-light but strong enough to survive the violent vibrations of a rocket launch—an acoustic environment equivalent to standing next to a jet engine at full power.

A Day in the Life: Three Perspectives

The daily life of a space engineer is a unique blend of deep, abstract analysis and meticulous, hands-on precision. While the satellite itself is thousands of miles away, the engineering work on the ground is a tangible and highly collaborative process. Your work will vary dramatically depending on your role and where the program is in its lifecycle, but it will almost always fall into one of three major categories: the digital world of design and analysis, the software world of command and control, or the physical world of building and testing.

The Thermal Engineer:

A mechanical engineer like "Alex" spends their week in a software world of pure physics, using a tool like Thermal Desktop. Their task is to build a high-fidelity thermal model of a new satellite, simulating how the sun's heat will be absorbed by different surfaces, how waste heat from the electronics will be conducted through the satellite's body, and how it can all be efficiently radiated back into the cold of space. A single error in their model, like a wrong surface property or a miscalculated conductance, could lead to the satellite overheating and dying months after launch, so every detail is scrutinized in intensive peer reviews.

The GNC Engineer:

An aerospace or software engineer's job is to ensure the satellite can point where it's supposed to. Their week is spent in MATLAB, writing and testing the Guidance, Navigation, and Control (GNC) algorithms. They might be modeling complex orbital maneuvers or testing the software that controls the satellite's tiny reaction wheels and thrusters. Their work is a delicate dance of control theory and orbital mechanics, ensuring that, millions of miles from Earth, the satellite can keep its antennas pointed at a ground station and its solar panels perfectly aimed at the sun with sub-millimeter precision.

The AIT Engineer:

An Assembly, Integration, and Test (AIT) engineer spends their time in the "clean room," a massive, pristine facility where the air is filtered to be cleaner than an operating room. Dressed head-to-toe in a "bunny suit" to prevent contamination, their week is a hands-on, meticulous process of physically building the satellite. Their ultimate responsibility culminates in the "shake and bake," a grueling test where the entire multi-ton satellite is mounted on a massive vibration table to simulate a rocket launch, and then placed in a giant thermal vacuum (TVAC) chamber to prove it will survive and operate in the harsh environment of space. Their motto is "test as you fly, fly as you test.."

What Makes The Space Domain Unique?

The Space Domain is governed by the absolute, unforgiving laws of orbital mechanics and the impossibility of repair. There are no second chances. This reality fosters a unique engineering culture, a mindset built on a foundation of extreme diligence and deep theoretical trust.

Reliability is a Religion: There is no higher virtue in space engineering than reliability because you only get one shot. If a satellite fails after launch, you cannot send a technician to fix it. This single, stark fact creates a culture of extreme diligence, of triple-checking every calculation, and of a testing regime more exhaustive than in any other domain. You will live by the mantra "test as you fly, fly as you test," meaning every component and the satellite as a whole must be proven on the ground to work in the vacuum and violence of space. Your work will be subjected to deep-dive peer reviews designed to find any and every potential flaw before it is enshrined in metal and silicon.

Analysis is King: In a world where you cannot build and break endless physical prototypes, the work is heavily dependent on high-fidelity modeling and simulation. Unlike a combat vehicle that can be run to failure on a proving ground, the cost and schedule for a satellite are so immense that the analysis must be right the first time. For a space engineer, the digital model is the design. A tremendous amount of trust is placed in the analysis, and your professional credibility will be built on your ability to create, run, and defend a complex simulation. The analysis report you write is not just a summary of your work; in the space world, it is as important as the hardware itself.

A Culture of Deep, Interdisciplinary Knowledge: While the Air Domain often fosters deep specialization, the sheer complexity and radical interconnectedness of a satellite requires its engineers to be more "T-shaped." No system operates in a vacuum. A thermal engineer, for instance, cannot simply analyze a "box;" they must have a deep understanding of the electrical systems inside it generating heat and the structural paths that conduct it. A GNC engineer must understand the structural dynamics and flexibility of the massive solar panels they are trying to point. The best space engineers are fluent in multiple technical languages, able to hold intelligent conversations with experts across the thermal, structural, electrical, and software disciplines.

Patience and Long Timelines: Satellites are among the most complex and reliable machines ever built, and that reliability is purchased with time. It is not uncommon for a major national security satellite to take five, ten, or even fifteen years to go from a concept on a whiteboard to the launchpad. This is not a world of rapid, six-month product cycles. Your career will be defined by patient, methodical, generational work, where you might spend a significant portion of your early career contributing to a single, monumental program. This requires a professional temperament that is comfortable with a long-term vision, where the ultimate reward is seeing a machine you helped build launched into orbit to perform its mission flawlessly for the next decade.

Key Roles & Major Players: Your Map of the Space Domain Ecosystem

The Space Domain is a deeply interdisciplinary field that demands a wide range of engineering talent. Building a system that can be launched on a controlled explosion, survive for decades in a vacuum, and perform a critical national security mission requires an incredibly diverse and collaborative team. While many specialties are needed, the core of any space program is built upon a foundation of key disciplines, each responsible for solving a fundamental piece of this extraordinary puzzle. Understanding these roles, and the major organizations that hire for them, is your first step in charting a course for a career in this final frontier.

Key Engineering Roles

While titles and responsibilities can vary from company to company, the work of a space engineer generally falls into a few critical categories. These are not just job descriptions; they are distinct disciplines, each requiring a unique mindset and a deep expertise in a specific area of physics or technology. A satellite is a symphony of these disciplines working in perfect harmony. These are the fundamental roles that turn a concept on a whiteboard into a functioning, billion-dollar asset in orbit, and excelling in any one of them can lead to a long and rewarding career as a sought-after expert.

Aerospace Engineers: These are the architects of the mission's journey and its life in orbit. They are the ones who answer the most fundamental questions: How do we get there, and how do we stay there? Aerospace engineers perform the complex astrodynamics calculations to design the trajectory that will take a satellite from the launchpad to its final, precise orbital slot 22,000 miles above the Earth. Once on orbit, their work shifts to Guidance, Navigation, and Control (GNC). They design and write the incredibly complex algorithms that act as the satellite's brain and nervous system, using data from star trackers and inertial sensors to control the tiny thrusters and internal reaction wheels that keep the satellite perfectly stable and its sensitive antennas and solar panels pointed with pinpoint accuracy. They live in a world of orbital perturbations, delta-v budgets, and advanced control theory.

Mechanical Engineers: A satellite is a delicate machine living in one of the most brutal environments imaginable, and mechanical engineers are responsible for its physical survival. They perform the deep structural analysis (FEA) to design a satellite "bus" (the main body) that is both feather-light and strong enough to survive the violent acoustic and vibrational loads of a rocket launch. Perhaps their most unique and challenging role is as thermal engineers. They are the masters of thermodynamics in a world with no air, designing the intricate systems of heat pipes, radiators, and reflective multi-layer insulation that keep the satellite's electronics from freezing in shadow or boiling to death in direct, unfiltered sunlight.

Electrical Engineers: These engineers design the "nervous system" and "circulatory system" of the satellite. They are responsible for the entire power subsystem, a miniature, self-sufficient power grid that must operate flawlessly for decades. This involves everything from the solar panels that collect energy to the batteries that store it and the sophisticated power distribution units that deliver clean, reliable voltage to every component. They also design the complex digital electronics and custom ASICs that serve as the "brains" of the spacecraft, often using specialized, expensive, and difficult-to-source radiation-hardened components that can withstand the constant bombardment of high-energy particles that would destroy conventional electronics.

Software Engineers: Modern satellites are sophisticated, software-driven robots, and the software engineers are their commanders. They write the millions of lines of ultra-reliable, real-time code that make up the flight software. This code commands every aspect of the satellite's operation, from firing a thruster to processing the data from its sensor payload. They also write the code for the ground systems that manage the communication links that send that precious data back to Earth. It is a world where a single bug cannot be patched with a simple download and could be mission-ending, demanding a level of rigor, testing, and fault tolerance that is far beyond that of typical commercial software.

The Dominant Commercial Players

The industrial landscape of space is a dynamic mix of legacy giants, who have the experience of building the nation's most critical assets, and new, disruptive players who are fundamentally changing the economics and the pace of the domain. On one side are the "Old Guard" Primes, companies with a deep heritage stretching back to the dawn of the Space Age. They are masters of the formal, rigorous systems engineering required for the nation's most sensitive and high-consequence missions. On the other side is the "New Space" movement, a vibrant ecosystem of innovative, often venture-backed companies that are bringing a faster, more agile, and commercially-driven mindset to the industry, challenging old assumptions and creating new possibilities.

The "Old Guard" Primes: Lockheed Martin Space, Northrop Grumman Space Systems, and Boeing's satellite division are the traditional titans of the industry. They have decades of proven heritage building the nation's most complex and high-consequence satellites, from the GPS constellation to the most secret reconnaissance assets. A career at one of these companies means working on some of the most significant and well-funded national security programs in history.

The "New Space" Innovators: The rise of companies like SpaceX, Blue Origin, and Rocket Lab has revolutionized the industry, primarily in the domain of launch. By pioneering reusable rockets, they have dramatically lowered the cost of getting to orbit, a paradigm shift that is enabling entirely new kinds of missions. While they are increasingly building their own massive satellite constellations (like Starlink), they also serve as critical partners, providing the launch services for many national security missions. Other new players, often backed by venture capital, focus on building smaller, more agile satellites at a faster pace.

The Government Customers & Research Centers

Unlike a commercial company that sells a product to the general public, the customer in national security space is a specific, mission-focused government organization. Understanding these key customers is essential because they define the "why" behind every line of code written and every piece of metal machined. The unique and demanding requirements of these organizations are what drive the incredible technological innovation in the industry. An engineer in this world is not just building a satellite; they are building a specific capability for a specific end-user, whether that's a Space Force Guardian managing a constellation, an NRO analyst reviewing imagery, or a soldier on the ground relying on a GPS signal.

The U.S. Space Force: As the newest branch of the military, the Space Force is the primary customer for and operator of military satellite systems. This includes the foundational constellations that the entire U.S. military depends on, such as GPS for navigation and timing, protected communications satellites, and missile warning systems. They define the requirements for these military-critical orbital assets.

The National Reconnaissance Office (NRO): Operating in deep secrecy, the NRO is the primary customer for the nation's intelligence satellites. They design, build, and operate the world's most advanced spy satellites, providing critical intelligence to the President and the broader Intelligence Community. For engineers who want to work on the absolute cutting edge of sensor, optical, and satellite technology, the NRO's programs are the ultimate destination.

Other Intelligence Agencies: Beyond the NRO, a number of other intelligence agencies are major consumers and drivers of space technology. They require specialized systems to meet their unique intelligence-gathering needs, creating a diverse and highly classified portfolio of projects across the industry.

The R&D Powerhouses: The Air Force Research Laboratory (AFRL), particularly its Space Vehicles Directorate at Kirtland Air Force Base, often pioneers the underlying, high-risk technologies that will be used in the next generation of satellites. They work in close partnership with NASA, which, while a civilian agency, often develops technologies (like new solar panels or communication systems for deep space probes) that are later adapted for national security missions due to their proven reliability in the harsh environment of space.

Engineering for the Final, Contested Frontier

To be an engineer in the Space Domain is to be an architect of systems that will outlive you. It is a world where your work, launched into the silent permanence of orbit, may continue to function for decades, which will be a distant and reliable monument to your engineering skill. The platforms you help design are the invisible infrastructure that runs our modern world and the vigilant sentinels that provide security from the ultimate high ground.

This culture is a direct result of the unforgiving environment and the impossibility of repair. The fanatical devotion to reliability, the deep trust in analysis, and the patient, long-term perspective are not choices; they are the hard-won lessons from the first sixty years of the Space Age. This is the "Space Domain Mindset."

But as you begin your career, you are stepping into a new and dynamic era. The final frontier is no longer a sanctuary. Your challenge, and the challenge of your generation of engineers, will be to invent the future of a resilient space architecture; a world that is defined by the need to operate in a contested domain, designing satellites that can maneuver and defend themselves, and creating vast, distributed constellations that can withstand losses. The Space Domain demands a unique blend of meticulous precision and bold innovation. This domain is for those engineers who thrive on complex physics and who wants to contribute to the most critical and rapidly evolving arena of the 21st century.

Chapter 11

The Cyber & Electronic Warfare Domain

In the conflicts of the 21st century, the most decisive battles may not be fought with steel, but with signals. Welcome to the invisible battlefield: the electromagnetic (EM) spectrum. Enter the domain of Cyber and Electronic Warfare (EW), a world of radar, radio, networks, and data where the goal is to achieve information dominance. Here, the battle is fought at the speed of light, where the terrain is made of frequencies and the weapons are algorithms. For the computer scientist or electrical engineer drawn to the fast-paced, problem-solving world of Silicon Valley, this domain offers something more: a chance to apply those same skills to challenges of unparalleled consequence, where the code you write and the circuits you design have a direct impact on national security.

The core principle is simple: see without being seen, hear without being heard, and connect without being detected. Engineers in this domain work to give friendly forces an overwhelming advantage within this invisible world, while denying that same advantage to the adversary. To understand this world, it's best to think of a simple analogy: a modern, networked computer.

A Legacy of Electronic Innovation

The history of this domain is a secret history of modern warfare, a "wizard war" fought in the shadows by engineers, physicists, and, later, by pioneering computer scientists. Unlike the saga of steel and gunpowder, this is a story of signals, codes, and algorithms, a relentless, cat-and-mouse cycle of innovation where a clever piece of hardware or a single line of code could be more decisive than an armored division.

The Wizard War Begins.

The story begins in the night skies over England during the Battle of Britain. The British development of radar was a revolutionary technological leap that allowed them to "see" incoming German bombers from miles away, a seemingly magical advantage. But the German response marked the true birth of Electronic Warfare. First came chaff, primitive strips of aluminum foil dropped from aircraft to create a blizzard of false targets on British radar screens.

Then came the more insidious battle of the beams. German bombers began navigating not by sight, but by following precise radio beams transmitted from stations in occupied France. The British response was not to try and jam these beams with brute force, but something far more clever: a technique called "meaconing." British engineers received the German signals and then re-broadcasted them, subtly bending the navigational beams to lure the German bombers into dropping their payloads over empty fields instead of London. This was not a battle of bombs, but a battle of wits fought in the invisible spectrum. It was the birth of deception jamming and the beginning of a relentless cycle: an advantage is gained, a countermeasure is developed, which requires a new countermeasure, and so on to this day.

The Cold War Listening Posts.

The Cold War turned this "wizard war" into a high-stakes, global enterprise. The primary mission became intelligence gathering. The U.S. and its allies deployed a massive, worldwide network of listening posts—from giant antenna farms in Turkey to reconnaissance aircraft like the RC-135 Rivet Joint and specially equipped submarines that would tap undersea cables—to hoover up every stray signal emanating from the Soviet Union. This was the dawn of Signals Intelligence (SIGINT).

This drove incredible innovation. It required the design of exquisitely sensitive receivers that could decipher the faint whispers of a Soviet radar from across the iron curtain, and the development of massive antenna arrays to pinpoint the source of a radio transmission. Most importantly, it spurred the first real-world applications of Digital Signal Processing (DSP). Early computer scientists and mathematicians were tasked with writing the complex algorithms needed to sift through mountains of intercepted static to find, classify, and decode a single valuable signal that could reveal a new Soviet military capability. It was the era of deep, patient, analytical work, where the hero was the analyst who could find the one crucial needle in the haystack.

The Digital Revolution.

The advent of the ARPANET—a DoD project designed to create a resilient, decentralized communications network—and its evolution into the Internet created a completely new, man-made battlefield: Cyberspace. The engineering challenge shifted in a fundamental way. The wizard war was no longer just about attacking and defending the radio waves carrying the data but about attacking and defending the very logic of the computers that processed the data.

The focus moved from jamming and deception to exploitation and intrusion. The weapon was no longer just radio frequency energy, but a malicious packet of data, a virus, or a "zero-day" exploit that could turn a computer against itself. As an engineer entering this domain today, you are stepping into this fully converged wizard war, where the line between a sophisticated Electronic Warfare attack and a cyber-attack has blurred completely, and the challenges are more complex than ever.

Mastering the Physics of the EM Spectrum

Think of Electronic Warfare as being primarily concerned with the physical layer of communication and sensing, the radio waves, infrared radiation, and other forms of electromagnetic energy that travel through the air and space. EW is the art and science of controlling this open-air medium. It is a world of applied physics, of antennas, of high-power transmitters and incredibly sensitive receivers. An EW engineer's job is to ensure that friendly forces can use the spectrum to their advantage while denying that same ability to the enemy. This timeless battle is broadly divided into three main functions:

Electronic Support (ES): Think of Electronic Support as the 'hearing' part of the wizard war; it forms the foundation of all electronic warfare. It is the art of building exquisitely sensitive receivers and antennas that can listen to the entire electromagnetic spectrum, often across vast distances, to detect the faint whispers of an adversary's activity. The engineering challenge is immense: to design a system that can pick a single, low-power enemy radio signal out of a sea of cosmic background noise, commercial radio stations, and friendly communications. Once a signal is detected, the challenge becomes one of classification. Is it a civilian air traffic control radar, or is it the targeting radar of a hostile surface-to-air missile system? Engineers in this space, often called SIGINT (Signals Intelligence) engineers, develop the complex algorithms and vast libraries of signal profiles needed to answer that question in a fraction of a second. This is the work that provides critical, life-saving intelligence.

Electronic Protection (EP): Electronic Protection functions as the 'shielding' part, the discipline of making friendly systems resilient to an enemy's electronic attack. It is the art of designing friendly communication and radar systems that can operate effectively even when an adversary is trying to jam them. This is not about brute force; it is about cleverness. Engineers in this space develop sophisticated techniques like frequency-hopping, where a radio rapidly jumps between thousands of different frequencies per second to evade a jammer. They design low-probability-of-intercept (LPI) radar waveforms that are so complex and faint, they are difficult for an enemy to even detect. This is a world of deep signal processing and clever hardware design; all focused on ensuring the link can never be broken.

Electronic Attack (EA): Conversely, Electronic Attack is the 'shouting' part, the active and aggressive side of EW. It is the art of transmitting powerful, precisely crafted radio energy to blind or deceive an enemy's electronic systems. This can be as simple as brute-force noise jamming, where you overwhelm an enemy radar with so much static that it cannot see its target. Or it can be as subtle and complex as deception jamming (spoofing), where you receive an enemy radar pulse, digitally alter it, and send back a false signal that makes their system think your aircraft is a hundred miles away from its actual location. This is a world of high-power RF engineering, sophisticated digital radio frequency memory (DRFM) circuits, and a deep, adversarial understanding of how an enemy's systems work.

Mastering the Logic of the Digital World

If EW is about the radio waves carrying the data, Cyber is concerned with the data itself and the computers that process it. Think of it as the world *inside* your laptop and the network it's connected to. Cyber warfare is not about jamming the Wi-Fi signal; it's about exploiting a flaw in the router's software to gain access to the network, or tricking a user into downloading malware that infects their computer. It is a world of pure logic, of code, of software vulnerabilities, and of network protocols. It is a man-made domain where the terrain is the architecture of a computer network.

Defensive Cyber Operations (DCO): The Art of the Digital Fortress. This is the "shielding" part for the digital world. It is the vast and unending task of protecting the DoD's immense global networks and, critically, the software embedded within its weapon systems. This involves far more than just running antivirus software. DCO engineers build complex firewalls and intrusion detection systems. They are software assurance experts who meticulously analyze the source code of a new missile's guidance system to find bugs before it's ever fielded. They are "Blue Teams," actively hunting for intruders within a network, and "Hunt Teams" that proactively search for vulnerabilities. Their work is the digital equivalent of armor plating, ensuring that a system can be trusted in a crisis.

Offensive Cyber Operations (OCO): The Art of the Digital Ghost. This is the "attack" part for the digital world, the work of finding and exploiting vulnerabilities in an adversary's networks and computer systems. This is a world of deep, forensic knowledge of how computer systems work, and a creative, out-of-the-box mindset for finding the one tiny flaw that allows access. It involves reverse engineering an adversary's software to understand how it works, developing custom tools to gain access, and navigating through a network without being detected. This is the work that allows for the disruption of an adversary's command and control or the gathering of critical intelligence without ever firing a shot.

The Convergence

Here is where it gets interesting, and where the most advanced engineering challenges lie. A modern weapon system, like an F-35, is not just a collection of hardware; it is a flying network of incredibly powerful, interconnected computers. This means that the line between EW and Cyber is blurring and, in many cases, has disappeared entirely. Is a sophisticated electronic attack that sends a precisely crafted false signal to a radar's receiver, tricking its software into thinking there's a target that isn't really there, an EW or a Cyber attack? The answer is both. Is an attack that exploits a vulnerability in the data link of an unmanned drone to take control of it an EW or a Cyber attack? Again, it's both. This convergence is the new reality. A modern engineer in this domain must appreciate both the physics of the electromagnetic spectrum and the deep, logical world of the software that runs on the systems that use it.

The Crown Jewel Programs

The programs in the Cyber and Electronic Warfare domain are often less visible than a massive aircraft carrier or a tank, but they are no less critical. They are the invisible systems that give every other platform its decisive edge. These programs are the embodiment of information dominance, focused on providing friendly forces with an unmatched understanding of the battlefield while blinding and confusing the adversary. The engineering here is a delicate and complex art, creating the "phantoms in the machine" that are the true force multipliers of modern warfare.

The Sentinels of the Spectrum

The foundation of information dominance is the ability to hear without being heard. These programs are the exquisitely sensitive ears that provide life-saving intelligence.

The F-35's Barracuda (AN/ASQ-239 System): This is the pinnacle of integrated, passive sensing. Developed by BAE Systems, the F-35's EW system is not just for self-defense; it is a powerful intelligence-gathering node. It can passively listen to the entire electromagnetic spectrum, identify enemy radar and communication signals, classify them, and even geolocate their source, painting a rich, real-time electronic picture for the pilot and sharing it with the rest of the fleet, all without ever emitting a signal of its own.

The Phantoms in the Machine

This is the active and aggressive side of the domain, focused on blinding, deceiving, and disrupting the adversary's digital nervous system.

The Navy's Next Generation Jammer (NGJ): This powerful and highly advanced pod, built by Raytheon, flies on the EA-18G Growler aircraft. Its purpose is to escort strike missions and blind enemy air defense radars with overwhelming, precisely targeted radio frequency energy. It is a masterpiece of RF engineering, using active electronically scanned arrays (AESA) and sophisticated software to generate incredibly complex jamming waveforms that can deceive even the most advanced radar systems.

The Digital Ghosts

This is the highly classified and deeply mysterious side of the Cyber Domain. If Defensive Cyber is about building an unbreakable fortress, Offensive Cyber Operations (OCO) are about finding the secret passage into the adversary's castle. The work is not about overwhelming a system with brute force; it is about the patient, forensic art of discovery and exploitation.

Cyber National Mission Force (CNMF): This is less a single "program" and more the operational heart of the U.S. Cyber Command. Engineers and operators on the CNMF work to find and exploit vulnerabilities in an adversary's networks and computer systems. This could involve reverse engineering an adversary's software to find a single flaw, developing custom tools and exploits to gain access, and navigating through a network without being detected. The goal is to gather critical intelligence or, if directed, to disrupt an adversary's command and control systems without ever firing a shot. A career in this world is a "digital chess match," requiring a creative, out-of-the-box mindset and a deeply intimate knowledge of how computer systems truly work.

The Digital Shields

While less glamorous, the most critical and enduring mission in this domain is protecting our own systems from attack.

Defensive Cyber Operations (DCO) for Weapon Systems: This is not a single program, but a broad and vital category of work across all domains. It involves the "Red Teams" of ethical hackers who are given a new weapon system like a tank or a missile and paid to find its digital vulnerabilities before an adversary does. It is also the work of the "Blue Teams," the software assurance engineers who meticulously analyze the source code of a system's flight-critical software to find and fix bugs before it is ever fielded. This is the world of penetration testing, reverse engineering, and digital forensics in a high-stakes environment.

Core Engineering Challenges

The engineering challenges of the Cyber and Electronic Warfare domain are a unique blend of deep physics, abstract mathematics, and a creative, adversarial mindset. Unlike a mechanical engineer who fights against gravity and stress, an engineer in this domain fights against the laws of information and the actions of intelligent adversaries. The work is a constant intellectual arms race, where a breakthrough in an algorithm or a newly discovered software vulnerability can be more decisive than a ton of armor. The problems you will solve are some of the most complex and intellectually demanding in the entire engineering profession.

Digital Signal Processing (DSP): This is the heart and soul of modern EW. It is the complex mathematics and algorithm development required to find a single, faint enemy signal in a sea of noise. As a DSP engineer, your work lives in MATLAB, Simulink, and often custom C++ or Python code. You will be designing the sophisticated digital filters, modulation/demodulation schemes, and Fast Fourier Transform (FFT) algorithms that allow a receiver to turn a chaotic stream of radio waves into intelligible data. A deep love for advanced math and a talent for algorithm design are essentially a requirement.

RF & Antenna Design: This is the hardware that touches the spectrum. An RF engineer's challenge is to design the antennas, transmitters, and receivers that can operate with incredible power, sensitivity, and precision. The cutting edge of this field is the Active Electronically Scanned Array (AESA) antenna. Unlike a traditional dish that has to be physically pointed, an AESA can steer a beam of energy electronically in microseconds, allowing it to track hundreds of targets or even perform multiple functions (like radar, electronic attack, and communications) simultaneously.

Cybersecurity & Network Hardening: This is the art of digital defense. For engineers in this space, the job is to think like a hacker and try to "break" your own systems. The work involves penetration testing (actively trying to find and exploit vulnerabilities), reverse engineering captured malware to understand how it works, and designing the secure network architectures and cryptographic systems that protect everything from a pilot's data link to a command center's internal network.

Algorithm Development & AI/ML: This is the new frontier where the future of the wizard war will be won. With the electromagnetic spectrum becoming more crowded and adversaries using more complex and novel signals, the future is in cognitive EW and cyber systems. The challenge for AI/ML engineers is to build systems that use machine learning to analyze the invisible battlefield in real-time, identify a new, never-before-seen enemy signal or cyber-attack, classify its intent, and automatically devise and deploy a novel countermeasure on the fly, all in a fraction of a second. This intellectually demanding field requires a deep understanding of both algorithm design and the domain's physics, creating a world where a clever algorithm can be the ultimate weapon.

A Day in the Life: Three Perspectives

The daily life of an engineer in this domain is often a world away from a manufacturing floor or a traditional hardware lab. It is a world of screens, of data, and of deep, focused thought. Much of the most critical work happens in secure, windowless facilities known as SCIFs (Sensitive Compartmented Information Facility), where engineers can work with the nation's most sensitive intelligence and capabilities. Your work will vary dramatically based on your specialty, but it will almost always involve a high-stakes, intellectual chess match.

The DSP Engineer:

An electrical or computer engineer might spend their week entirely within MATLAB and Simulink, completely removed from the hardware. Their mission is to design a new filtering algorithm to pull a faint, previously undetectable low probability-of-intercept (LPI) enemy signal out of a dense signal environment. This is not just an academic exercise; success means giving a submarine commander a critical piece of intelligence that prevents a surprise attack. Their world is one of pure mathematics and logic, a cycle of simulating the algorithm, testing it against recorded real-world signals, and tweaking it to achieve the perfect balance of sensitivity and noise rejection. Much of their work is done in a SCIF, analyzing highly classified signal data.

The RF Engineer:

This engineer lives in a lab that looks like a fortress of black foam cones: the anechoic chamber. Their job is to translate the digital world into the physical world of electromagnetic waves. Their week might be spent meticulously testing a new AESA antenna design. Using tools like a vector network analyzer, they are in a hands-on dialogue with physics, measuring the power, beam width, and sidelobe performance of their hardware. This work is critical because a single decibel of improved efficiency in an antenna could mean a jammer can protect an entire squadron of aircraft instead of just one.

The Cyber Operator:

A computer scientist's "lab" is a keyboard. Their week is spent in a secure facility, acting as a member of a "Red Team," a group of ethical hackers paid to think like the enemy. Their job is to find a way into a new weapon system before an adversary does. They might be writing Python scripts to probe a tank's network for vulnerabilities or using a disassembler like Ghidra to reverse-engineer a drone's firmware, looking for a single exploitable flaw. Their importance cannot be overstated; they are the digital immune system for the entire military, finding and fixing the vulnerabilities that could otherwise be catastrophic in a crisis.

What Makes The Cyber & EW Unique?

This domain has no single physical tyrant like gravity or pressure. It is governed by a more cunning and unpredictable force: the intelligent adversary. Every design choice, every line of code, and every waveform is created with the full knowledge that a determined, creative human being on the other side is actively trying to defeat it. This fosters a unique engineering culture of paranoia, creativity, and relentless, real-time adaptation.

The Ultimate Cross-Cutting Domain: While we've discussed the Air, Land, Sea, and Space domains as separate arenas, the Cyber and EW domain cuts across and influences all of them. A ship at sea is protected by an EW system. A tank on the ground relies on a secure data link. A satellite in orbit is a potential cyber target. This makes the Cyber/EW domain the connective tissue, the nervous system that ties the entire multi-domain battlefield together. An engineer here must be able to think about how their system will function in every other environment.

Think Like the Adversary: More than in any other domain, engineers here must constantly put themselves in the shoes of a determined, intelligent adversary. You must be paranoid. How could this signal be intercepted? How could this network be breached? This "Red Team" mindset of always looking for your own weaknesses is a core cultural value.

Software is the Weapon: While hardware is the enabler, the true capability of a modern EW or Cyber system lies in its software. A new line of code can create a new jamming waveform that can make a billion-dollar radar system go blind, or a new cyber defense that can protect an entire carrier strike group. This creates a culture of rapid, iterative software development, where a clever algorithm can be more decisive than a ton of armor.

The Need for Secrecy: Due to the incredible sensitivity of these capabilities, much of the work in this domain is highly classified. This fosters a culture of discipline and discretion. Engineers work in secure facilities (SCIFs) and cannot discuss the details of their work, even with their families.

Key Roles & Major Players

The engineers who thrive in the Cyber and Electronic Warfare domain are a special breed. They are the abstract thinkers, the puzzle solvers, and the digital detectives who are comfortable operating in a world of pure information. A strong foundation in the underlying mathematics and physics is non-negotiable, as is a passion for the relentless, cat-and-mouse game of technological competition.

Engineers in High Demand

While titles and responsibilities vary from company to company, the work of an engineer in this domain generally falls into a few critical categories. These are not just job descriptions; they are distinct disciplines, each requiring a unique mindset and a deep expertise in a specific area of physics or technology. A modern EW system is a symphony of these disciplines working in perfect harmony, and excelling in any one of them can lead to a long and rewarding career as a sought-after expert.

Electrical Engineers: EEs are the foundational talent for this domain, particularly for Electronic Warfare. They are the ones who understand the deep physics of electromagnetic wave propagation, antenna theory, and radio frequency (RF) circuit design. They are in high demand because they can build the hardware that touches the spectrum: the high-power transmitters, the exquisitely sensitive receivers, and the custom microelectronics (ASICs and FPGAs) that can process signals at billions of samples per second. An EE in this field is a true applied physicist, turning Maxwell's equations into a tangible capability.

Computer Engineers: CEs are the critical bridge between the physical world of electronics and the logical world of software. They have a unique, hybrid skillset that is perfectly suited to the converged nature of modern Cyber and EW. They are in high demand because they can work at the lowest levels of a system. They write the firmware; the software that lives directly on the microchips. They understand the deep architecture of a processor and can perform the reverse engineering on a piece of hardware to find its vulnerabilities.

Computer Scientists: CS graduates are the masters of the purely digital realm. They are in high demand to write the high-level Digital Signal

Processing (DSP) algorithms that filter and classify signals. They are the primary talent for Defensive and Offensive Cyber Operations, building the firewalls, intrusion detection systems, and the sophisticated tools used to analyze and exploit network vulnerabilities. With the rise of the Third Offset Strategy, their skills in Artificial Intelligence and Machine Learning are now perhaps the single most sought-after talent in the entire defense industry.

The Dominant Commercial Players

While many companies operate in this space, the landscape of high-end Electronic Warfare and Signals Intelligence is dominated by four undisputed leaders. These companies have decades of experience, deep classified knowledge, and massive R&D budgets dedicated to mastering the invisible battlefield.

Raytheon (an RTX Company): Often considered the world's premier radar and sensor house, Raytheon's expertise in this domain is legendary. They build the massive AESA radars for platforms like the F-15 and F/A-18, the advanced sensors for spy satellites, and are the creators of the Next Generation Jammer (NGJ), one of the most sophisticated electronic attack systems ever built. A career here is often a deep dive into the physics of RF and the hardware that generates and receives electromagnetic energy.

Northrop Grumman: A powerhouse in airborne EW and battlefield communications. They are experts in building the complex, integrated suites that protect aircraft from missile threats and are leaders in developing the resilient data links and networking technologies that are the backbone of modern C4ISR. They are masters of creating systems that can operate in heavily jammed and contested environments.

BAE Systems: With a deep heritage in electronic warfare, BAE Systems is a world leader in defensive systems and signals intelligence. They are the creators of the F-35's incredibly complex AN/ASQ-239 "Barracuda" system, which provides the pilot with an unparalleled, 360-degree view of the electronic battlefield. They are experts in the art of listening, building the sensitive receivers and algorithms that can detect and identify threats.

L3Harris Technologies: A giant in the world of communications and networking. L3Harris is a leader in building the secure, encrypted radios and data links that soldiers, sailors, and airmen use to communicate. They are experts in creating resilient waveforms and networking protocols that ensure the lines of communication stay open, even under direct attack.

The Government Customers & Research Centers

The government's role in this domain is twofold: it is the ultimate customer that defines the requirements, and it is also a primary hub of research and employment for engineers.

The Customers: The various branches of U.S. Cyber Command (CYBERCOM) are the primary operators and drivers of requirements for both defensive and offensive cyber operations. In the world of signals intelligence, the undisputed center of gravity is the National Security Agency (NSA), located at Fort Meade, Maryland. The NSA is one of the largest single employers of mathematicians and computer scientists in the world, and it is the nation's premier organization for both collecting foreign intelligence and protecting U.S. national security systems. Other intelligence agencies are also major customers for specialized capabilities.

The Research Centers: The service research labs are where the foundational, high-risk R&D takes place. The Air Force Research Laboratory (AFRL) has directorates focused on sensors and information technology that pioneer the next generation of EW and C4ISR concepts. The Naval Research Laboratory (NRL) in Washington, D.C., has a deep history of innovation in this area, having done foundational work in radar and space-based reconnaissance. For an engineer who wants to live in the world of pure R&D, these government labs are a top destination.

The Wizard War

To be an engineer in the Cyber and Electronic Warfare Domain is to be a modern-day wizard, fighting a war of wits on an invisible battlefield. It is a world of abstract puzzles, complex algorithms, and high-stakes digital chess, where the goal is to outthink and outmaneuver the adversary. The systems you build are the shields that protect the nation's networks, the eyes that can see in the dark, and the voice that can whisper across continents.

This culture is a direct result of its unique role as the great integrator. The deep respect for fundamental physics, the adversarial mindset, and the understanding that software is the weapon are the core tenets of this Wizard War. As you begin your career, you are stepping onto a battlefield that is constantly changing and touches every other domain. Your challenge will be to invent the future of information warfare: to build cognitive, AI-driven systems that can adapt to new threats in real-time and to create unbreachable cyber defenses that will protect the next generation of military platforms. For the engineer who thrives on deep intellectual challenges and who wants to operate at the very heart of modern, multi-domain conflict, there is no more dynamic or critical field.

Part 3

Technology, Strategy, and Your Place in Its Future

Chapter 12

The Engineer of 2040

You now have a detailed map of the defense industry as it exists to-day. You understand its culture, its domains, and its key players. But a map of the present is not a guide to the future. The technologies that will define the middle of your career, the challenges you will be asked to solve in the 2030s and 2040s, are being invented in laboratories and research centers right now. To build a truly resilient and successful long-term, future-proof career, it is not enough to have the skills that are in demand today. You must cultivate the habit of strategic foresight.

Strategic foresight is the ability to look at the horizon, to identify the disruptive technological and geopolitical trends that are gathering momentum, and to strategically invest in the skills that will be valuable tomorrow. This is the essential mindset that separates the engineer who reacts to change from the one who architects a future-proof career. It is the difference between passively letting your career happen *to* you and actively designing a career that will keep you relevant and in high demand for decades. It is the most important mindset for ensuring your longevity and success in a field defined by rapid technological change.

This chapter is your first strategic intelligence briefing. We will take a deep dive into five technological and strategic megatrends that are fundamentally reshaping the defense landscape. Understanding these trends will help you make smarter decisions about the training you seek, the master's degree you pursue, and the career path you choose to follow, ensuring you are not just prepared for your first job, but for your entire career.

The Primacy of Software, AI, and the Data-Driven Battlefield

For the entirety of the 20th century, the dominant paradigm of military power was platform centric. The nation with the best hardware, the fastest jet, the thickest tank armor, the quietest submarine, held the advantage. Engineering was a tangible, physical discipline focused on optimizing these incredible machines. While computers were involved, they were almost always in a supporting role, subservient to the physical platform they inhabited. That entire paradigm has now been inverted.

Welcome to the defining megatrend of the 21st century: the primacy of software. Today, the hardware is increasingly becoming a vessel for the software that is the true source of its capability. An F-35 is not a jet that happens to have a computer; it is a flying supercomputer that happens to have a jet engine. Its true power comes not from its speed, but from the millions of lines of code that fuse its sensor data, manage its electronic warfare suite, and network it with the rest of the force. This is a fundamental shift in the very nature of engineering, and it is the single most important trend for a young engineer to understand.

Victory on the future battlefield will not go to the side with the most platforms, but to the side with the fastest and best decisions. It will be a torrent of data, flowing from satellites, drones, ground sensors, and soldiers' headsets. The victor will be the one who can process that data, use artificial intelligence to identify patterns and targets, and get that information to the right person or system in a fraction of a second. Artificial Intelligence and Machine Learning are no longer just buzzwords; they are the core technologies of the Third Offset Strategy, the tools that will provide the next great leap in military advantage.

The Rise of the Software-Defined Platform

The term "software-defined" is a powerful concept that a young engineer must grasp, as it represents a fundamental shift in the philosophy of engineering. It means that a platform's core functions are no longer permanently locked into the physics of its hardware but are instead controlled by flexible, adaptable, and endlessly updatable software. This transition from a hardware-centric to a software-centric mindset is happening across every domain and is changing the very nature of what an "upgrade" means.

Consider the evolution of a radio. A traditional radio's capabilities were determined by its physical, analog circuits, its filters, its mixers, and its amplifiers were all carefully crafted pieces of hardware. To change its function, to add a new frequency band or a new modulation scheme, you had to physically rip out the old hardware and replace it with a new circuit board. A modern software-defined radio (SDR), in contrast, is a piece of generalized hardware (an antenna and a fast analog-to-digital converter) that is controlled by powerful software. By simply changing the code, that same piece of hardware can become a GPS receiver, a cellular radio, a secure military communications device, or even a signals intelligence collection system. The hardware is just the portal to the spectrum; the software defines what it does.

This trend is now happening to everything. A modern AESA radar system's capabilities are defined not only by its antenna, but also by the complex signal processing algorithms that can sift through noise, detect a stealthy target, and even perform electronic attack functions. A modern jet engine's performance is managed by its FADEC (Full Authority Digital Engine Control), a computer whose software can be updated to improve fuel efficiency or increase power output without ever touching a turbine blade. This profound shift dictates that the value and capability of a multi-billion dollar platform can be dramatically increased not by a costly, time-consuming hardware replacement, but by a simple software update, a process that is orders of magnitude faster and cheaper.

This new reality has massive implications for the engineering lifecycle. It creates a culture of continuous improvement, where a system's capabilities are not fixed on the day it leaves the factory but are expected to evolve and improve over its entire 30-year service life. This is the world of DevSecOps (Development, Security, and Operations), an agile software development philosophy that is rapidly being adopted by the defense industry to push new capabilities from the lab to the warfighter in a matter of weeks, not years. It means that the software engineer is no longer a support player; they are a central driver of the platform's long-term relevance and lethality.

Your career, regardless of your major, will be inextricably linked to software. The lines between disciplines are blurring. You must become computationally literate. For a mechanical or aerospace engineer, this means learning to code in a language like Python. You will not be expected to write the flight-critical code, but you will be expected to write scripts to automate your analyses, to process the vast amounts of data from a modern test, and to interact with the complex digital models of the systems you design. The mechanical engineer who can script and analyze data is exponentially more effective and valuable than one who cannot. They can do in an afternoon what might have taken a non-coding engineer a week of manual spreadsheet work.

Artificial Intelligence and the Autonomous Revolution

Artificial Intelligence and Machine Learning are the engines of the next great technological leap. In the defense context, this is not about creating a "Terminator" or a sentient, self-aware AI. The reality is both more practical and more profound. It's about using computers to perform specific, difficult tasks faster, more accurately, and at a greater scale than any human can. This application of AI is what enables the rise of true autonomy, moving beyond simple automation (a system that follows a pre-programmed script) to a state where a system can perceive its environment, orient itself, and make decisions to achieve a goal, even in a novel situation.

The goal of this revolution is not to replace the human, but to empower the human. It is about augmenting human capability. In the near future, the goal is to allow a single operator to manage a team of ten robotic systems, or to have an AI co-pilot that can manage a fighter jet's sensors and defensive systems, freeing the human pilot to focus on the high-level tactical decisions. It is about using AI to help an intelligence analyst find the needle in the haystack, the one critical piece of information in a mountain of satellite imagery or a torrent of signals intelligence.

The core technologies driving this are often computer vision and deep learning. A computer vision algorithm, for example, can be trained on millions of images to automatically detect and classify a specific type of vehicle from the video feed of a drone, a task that would be mind-numbingly tedious for a human operator. A deep learning model can analyze the complex radio frequency signals from an enemy radar and identify its type in a fraction of a second, providing instant situational awareness. This is not just a faster way of doing old tasks; it enables entirely new capabilities and a new tempo of operations.

This pursuit also involves solving some of the hardest problems in AI research. For example, how do you train an AI to be effective in a data-scarce environment? Unlike a commercial AI that can be trained on billions of images from the internet, a military AI might need to be trained to identify a new, rare enemy system from only a handful of examples. This requires the development of new techniques like one-shot learning (the ability for an AI to learn from a single example) and the use of high-fidelity synthetic data. Furthermore, the AI must be robust against adversarial attacks, where an enemy tries to deliberately fool the system with confusing or deceptive inputs.

A foundational understanding of AI/ML concepts is becoming a core competency for all engineers. For Electrical and Computer Engineers, this is your new heartland. Your skills in signal processing and algorithm development are the raw material of the AI revolution. For Computer Scientists, this is your chance to work on some of the most challenging and meaningful AI problems in the world, where the stakes are infinitely higher than optimizing an ad network. For all disciplines, it is crucial to understand the language of AI. You should understand the difference between training data and inference, what a neural network is, and, most importantly, the limitations of these systems. The ability to have an intelligent conversation with the AI/ML experts will be a critical skill for any systems engineer in the coming decade.

The Data-Driven Battlefield and the Combat Cloud

The ultimate goal of this software and AI revolution is to create a fully networked, data-driven force. The concept is known as Joint All-Domain Command and Control (JADC2), or more simply, the "combat cloud." The vision is to create a military internet-of-things (IoT), where every platform, every ship, tank, plane, and soldier, is a sensor and a node in a massive, resilient, self-healing network. In this world, a sensor on an F-35 could detect a threat and seamlessly pass the targeting data to a Navy destroyer, which could then launch a missile to intercept it, all without a human ever acting as a manual switchboard operator. This compresses the sensor-to-shooter timeline from minutes or hours down to seconds, creating a tempo of operations that is impossible for a less-networked adversary to match.

This shift creates an entirely new set of engineering challenges. Rather than building one platform, you are building the network that connects them all. The problems are not just in getting the data from point A to point B, but in ensuring that the data is secure, that the network is resilient to attack, and that the right information gets to the right person or system at the right time. This is a system-of-systems problem of almost unimaginable complexity.

The engineering of this combat cloud requires a deep understanding of modern networking technologies, from the high-bandwidth links of 5G to the resilient, ad-hoc networking protocols that allow a swarm of drones to create their own network on the fly. It is a world of data fusion, where sophisticated algorithms must take in conflicting data from dozens of different sensors and create a single, coherent picture of the battlefield. It is a world where cybersecurity is not an afterthought, but a foundational requirement, as the network itself becomes a primary target for the enemy.

The future of engineering is interdisciplinary and network-focused. The value of your work will increasingly be measured by how well your component or system communicates and integrates with the larger whole. You must learn to think about interfaces, data formats, and network protocols. Understanding the basics of cybersecurity is no longer an optional skill for a software engineer; it's a mandatory skill for every engineer. A mechanical engineer designing a new sensor must now ask, "How will this sensor's data be securely transmitted, and in what format?" An electrical engineer designing a new radio must ask, "How will this radio join and leave a mobile, ad-hoc network?" This is the new reality of the data-driven battlefield.

Augmenting Human Intelligence

Beyond controlling platforms, one of the most profound impacts of AI is its role as a tireless, superhuman intelligence analyst. In the modern world, the challenge is not a lack of data, but a crushing overabundance of it. A single reconnaissance satellite can generate terabytes of imagery in a single day. A signals intelligence platform can collect millions of electronic emissions. It is impossible for a team of human analysts to review all of this data in a timely manner. This is where AI becomes a mission-critical force multiplier.

AI, particularly computer vision and natural language processing, is being developed to perform the initial, massive-scale analysis. An AI can be trained to scan thousands of square miles of satellite imagery to find every vehicle that matches the signature of a specific missile launcher. It can listen to millions of hours of communications to flag a conversation that uses a specific keyword. The AI doesn't replace the human; it acts as the world's most powerful filter, finding the handful of needles in the haystack and presenting them to a human analyst for the final, critical judgment.

This same principle applies at the tactical edge. In the cockpit of a future fighter, an AI co-pilot will be able to analyze the entire electromagnetic spectrum, identify the most dangerous threats, and instantly suggest the optimal countermeasure or evasion tactic to the human pilot, allowing the pilot to make a life-or-death decision in a fraction of the time. In the world of augmented intelligence, the goal is to fuse the raw processing power of the machine with the intuition and ethical judgment of the human operator.

The demand for computer scientists and data scientists who can build, train, and validate these AI models is exploding. For all engineers, it means understanding that the systems you build will be part of this data ecosystem. The sensor you design must produce data that is clean and easily digestible by an AI. The user interface you create must be able to clearly and intuitively present the output of an AI's analysis to a human operator. Data fluency is no longer optional.

The Digital Twin Revolution and Model-Based Systems Engineering (MBSE)

For the vast majority of engineering history, the "truth" of a design was captured in a physical document. From the hand-drawn vellum blueprints of a World War II battleship to the 2D CAD drawings of a Cold War fighter jet, the design was communicated through a series of disconnected, static, and often out-of-date papers, spreadsheets, and reports. This was an inefficient and incredibly error-prone system. An engineer in the structures department might be working from a version of a drawing that was two weeks old, unaware that a change made by the propulsion team had just rendered their entire analysis invalid. This document-based approach was the source of countless costly mistakes and debilitating delays.

That entire paradigm, and the culture of rework and inefficiency it created, is being completely replaced by a new philosophy of engineering. Welcome to the second megatrend: the "Digital Twin" revolution. This is the industry-wide shift toward Model-Based Systems Engineering (MBSE), a world where a single, high-fidelity, interconnected digital model of the entire system, the digital twin, is the one and only source of truth. This model is born at the very beginning of the program and lives, breathes, and evolves with the physical hardware from the initial concept through manufacturing, sustainment, and eventual retirement.

In this new world, the model is not just a picture of the design; the model *is* the design. The requirements are not in a separate document; they are linked directly to the digital components. The analysis models are not built from scratch; they are directly driven by the authoritative design geometry. The manufacturing instructions are not interpreted from a 2D drawing; they are generated directly from the 3D model's embedded information. This digital thread connects every single aspect of the program, creating a far more efficient, accurate, and collaborative environment.

The Digital Twin and the Authoritative Source of Truth

The concept of the Digital Twin is at the heart of this revolution. It is the idea that for every physical platform that exists in the real world—every jet, every ship, every satellite—there is a corresponding, perfectly mirrored digital version that lives on the program's servers. This is not just a 3D CAD model; it is a rich, multi-dimensional, and dynamic representation of the system. The digital twin contains not only the geometry of every part, but also the material it's made from, the supplier it comes from, the requirements it's meant to verify, the analysis reports associated with it, the software that runs on it, and eventually, the real-world performance and maintenance data streamed back from the physical asset once it's fielded.

This creates an "authoritative source of truth." In the old world, truth was fragmented across a dozen different documents, and a change in one place (like a requirements document) required a massive, manual effort to update every other document. The digital twin solves this. When an engineer needs to know the exact weight of a component, they don't look at a spreadsheet that might be out of date; they query the digital twin. When a manufacturing engineer needs to know the tolerance on a specific hole, they don't look at a drawing; they query the digital twin. When a software engineer needs to know the power consumption of a circuit board, they query the digital twin.

This paradigm shift has profound implications. It enables virtual integration, where engineers can assemble and test the entire system in a digital environment, finding and fixing integration errors long before any metal is cut. It allows for predictive maintenance, where sensor data from a real-world jet engine can be fed back into its digital twin to predict when a part will fail. It creates a level of shared understanding and collaboration across vast, geographically dispersed teams that was previously impossible.

You must develop your understand of digital tools and the philosophy of digital engineering. Your value is increasingly tied to your ability to create, interpret, and collaborate within this fully digital environment. You must treat the digital model with the same respect and discipline as the physical hardware. An error in the digital twin is not just a typo; it is a fundamental design flaw that will eventually become a very expensive physical problem.

Model-Based Systems Engineering (MBSE) and the Language of Systems

If the digital twin is the "what," Model-Based Systems Engineering (MBSE) is the "how." MBSE is the formal, structured methodology for creating and managing the digital twin, particularly at the front-end of the design process where the system's requirements and architecture are defined. A key component of MBSE is the use of a standardized, graphical modeling language, the most common of which is called SysML (Systems Modeling Language). SysML is to a systems engineer what C++ is to a software engineer or what GD&T is to a mechanical engineer: it is the formal, unambiguous language of their profession.

Using a tool like Cameo Systems Modeler, an engineer uses SysML to create a series of interconnected diagrams that capture every aspect of the system's function, form, and behavior. They create requirements diagrams to capture what the system must do. They create block definition diagrams to show how the system is broken down into its major components. They create activity diagrams to model the system's behavior. The power of MBSE is that all these diagrams are part of a single, underlying model. They are not just pretty pictures; they are different views of the same authoritative data. A change in a single requirement will automatically flag every single part of the design, every interface, and every test case that is affected by that change.

You must embrace the mindset of systems thinking. Even as a component designer, you are a node in this larger digital model. You should learn the basics of MBSE and SysML if you ever get the chance, as it is rapidly becoming the dominant language for how complex systems are designed and managed. Understanding how to create and manage requirements in a digital, model-based environment is a massive advantage that will make you a highly sought-after systems engineer.

The Digital Thread and the Future of Manufacturing

The "Digital Thread" is the practical result of the MBSE philosophy. It is the seamless, interconnected flow of data from the initial digital concept all the way to the physical product on the manufacturing floor, and beyond into its operational life. In the old world, data had to be manually re-entered or translated between different systems (from design to analysis, from analysis to manufacturing), a process that was slow and full of errors. The digital thread automates this.

For example, a mechanical engineer designs a bracket in Creo. That same digital model is then seamlessly imported into ANSYS for stress analysis. Once approved, the model, with its embedded Product and Manufacturing Information (PMI) and GD&T, is sent directly to the Computer-Aided Manufacturing (CAM) software that generates the toolpaths for the CNC machine that will cut the part. The same model is then used by the quality inspector with a digital CMM (Coordinate Measuring Machine) to verify the final part's dimensions against the original digital authority.

You must become a power user of your core software and understand how it connects to the other parts of the engineering lifecycle. Do not be content with just knowing the basics of your CAD tool. Pursue advanced certifications (like the CSWP for SOLIDWORKS). Learn about Product Data Management (PDM) and Product Lifecycle Management (PLM) systems like Windchill or Teamcenter, as these are the software backbones that manage the digital thread. The engineer who understands not just how to design the part, but how that digital design will be used by the analysts and the manufacturers, is an invaluable asset in the age of the digital twin.

The Rise of Networked, Autonomous, and Collaborative Systems

The era of the single, exquisite, and incredibly expensive manned platform is evolving. While these systems will remain critical, the future of the force lies in distributed, collaborative systems, swarms of lower-cost, unmanned assets that work together as a team. A future air battle might not be a handful of manned fighters, but two manned aircraft commanding a swarm of a hundred autonomous drones. A future naval fleet might consist of a few manned warships commanding a flotilla of unmanned surface and undersea vessels. This is the concept of manned-unmanned teaming (MUM-T), and it is a central pillar of future defense strategy.

This creates a paradigm shift in the engineering challenges. The focus moves from optimizing the performance of a single platform to optimizing the performance of the *entire network*. The lethality and survivability of the system is no longer just in the platform, but in the resilience and speed of the data links that connect them, and the intelligence of the algorithms that manage them as a cohesive team.

Manned-Unmanned Teaming and the Loyal Wingman

Manned-Unmanned Teaming (MUM-T) is the concept of creating a symbiotic, collaborative relationship between human operators and autonomous systems, a cornerstone of the Third Offset Strategy. The human is not a remote-control pilot, painstakingly flying a single drone with a joystick. Instead, the human acts as a mission commander, a high-level strategist who commands a team of intelligent, autonomous agents. The autonomous systems act as their loyal, intelligent wingmen, sent ahead to perform the "dull, dirty, and dangerous" tasks, reducing the risk to human life.

A prime example of this is the "Loyal Wingman" concept, programs like the Air Force's Collaborative Combat Aircraft (CCA). The vision is for a single manned fighter jet to be accompanied into combat by a team of five or ten unmanned, AI-driven drones. These CCAs would fly ahead of the manned jet, using their own sensors to scan for threats and identify targets. They could act as decoys, as electronic jammers, or even as weapons platforms, all under the strategic direction of the human pilot safely in the rear.

This concept extends to every domain. In the Land Domain, it's a manned tank commanding a team of Robotic Combat Vehicles. In the Sea Domain, it's a destroyer acting as the mothership for a fleet of unmanned undersea vehicles. This requires not just brilliant robotics, but a deep understanding of human factors engineering to create the intuitive command and control interfaces that make this complex symphony possible. It also demands a new level of networking, where systems like advanced Blue Force Tracking do not just show where friendly humans are, but where every single autonomous asset is on the battlefield, creating a complete, real-time picture for the commander.

This is the ultimate interdisciplinary field. It requires mechanical engineers to design the robotic platforms, electrical engineers to design their sensors and communication systems, software engineers to write the control and AI algorithms, and human factors engineers to design the interface between the human and the machine. If you are a "T-shaped" engineer who is comfortable working across disciplines and who is fascinated by the intersection of human psychology and robotics, this is the field for you.

The Challenge of the Resilient Network

For a swarm of autonomous systems to work together as a team, they must be able to communicate. This creates one of the most difficult and critical engineering challenges of the modern era: how do you build a resilient, mobile, ad-hoc network that can function in a chaotic and hostile electromagnetic environment? An adversary will be actively trying to jam your data links, spoof your GPS signals, and hack into your network. A swarm of brilliant robots that cannot talk to each other is just a collection of useless, isolated machines.

The solution is not a single, perfect radio, but a diverse and multi-layered web of communication technologies. This includes everything from high-bandwidth, directional line-of-sight data links for short-range communication, to jam-resistant, frequency-hopping satellite communications for over-the-horizon links, and even experimental optical (laser-based) communication systems that are nearly impossible to intercept. The engineering challenge is not just in the hardware, but in the networking software. These are the sophisticated protocols that allow the swarm to form its own network on the fly, to "heal" itself if a node is lost or jammed, and to intelligently route data through the most secure and reliable path available at any given moment.

RF engineering is a world of complex waveform design, and of high-level network architecture. The goal is to create a "nervous system" for the force that is as resilient and adaptable as a biological organism, one that can continue to function even when it is damaged.

The ability to design systems that can communicate reliably and securely is becoming more important than optimizing the performance of any single component. Electrical and computer engineers with a deep understanding of RF communication, networking protocols, information theory, and cybersecurity are in incredibly high demand. For all engineers, it means you must design your systems with the network in mind from day one. Your component is not a standalone object; it is a node on a network, and its ability to communicate is a primary requirement.

The Ethics and Trust of True Autonomy

As autonomous systems become more capable and are given more independence, the industry is grappling with a profound set of ethical and engineering challenges centered on a single, critical concept: trust. A human commander must be able to trust that an autonomous system will perform its mission reliably, that it will not make catastrophic errors, and, most importantly, that it will operate within the strict, established legal and ethical boundaries known as the rules of engagement. This is particularly critical as we approach the possibility of systems that are authorized to use lethal force.

This is not just a philosophical problem for policymakers; it is a deep and difficult engineering challenge. How do you design an AI so that its decision-making process is not an inscrutable "black box"? How do you perform Verification and Validation (V&V) on a complex, deep-learning system whose behavior may not be perfectly predictable in every single edge case? How do you build a system that can be mathematically proven to be "safe"? This is the frontier of Explainable AI (XAI) and AI safety and ethics.

The engineering work in this area is some of the most important in the world. It involves developing new methods for testing and evaluating AI systems. It requires the creation of sophisticated simulation environments where an AI can be run through millions of virtual scenarios to find its potential failure modes. It involves designing the AI's logic in such a way that it can later explain *why* it made a particular decision, providing a crucial audit trail for human commanders.

The demand for engineers who can think not just about capabilities, but also about safety, reliability, testing, and ethics, will grow exponentially. This is a field that requires a deep sense of responsibility and a mature, thoughtful approach. If you are interested in the intersection of technology and ethics, and if you want to work on the problems that will define the safe and responsible use of AI for decades to come, this is your calling. A career in test and evaluation (T&E) or systems safety engineering for autonomous systems is a challenging and incredibly high-value path.

The Rise of Cheap, Precise Munitions and the New Defense

The nature of warfare is being democratized by technology. The proliferation of cheap, highly effective, and often commercially-derived systems, like small "suicide drones," has created a new and complex threat. These systems, which can be purchased for thousands of dollars instead of millions, allow smaller actors to threaten incredibly expensive assets like an aircraft carrier or an air base. This has created an urgent engineering challenge: how do you affordably defend against a swarm of a hundred cheap drones? Firing a million-dollar missile to shoot down a thousand-dollar drone is not a sustainable solution.

This has led to a massive investment in Counter-Unmanned Aerial Systems (C-UAS) technology. This is a fascinating and rapidly evolving field. It includes the development of new, smaller radar systems that can detect and track these tiny targets. It involves creating sophisticated electronic warfare systems that can jam their control links or spoof their GPS signals, causing them to fall out of the sky. And it is a primary driver for the Directed Energy weapons we discussed earlier, as a high-energy laser offers the promise of a near-infinite magazine with a very low cost-per-shot, a perfect solution for the drone swarm problem.

This is a world of rapid prototyping and intense, real-world feedback. The threats are evolving every month, and the defensive systems must evolve even faster. It is a domain that rewards clever, interdisciplinary thinking and the ability to create a solution that is not just effective, but also affordable at scale.

This is a wide-open field for innovation. It requires RF engineers to design the new sensors and jammers, software engineers to write the tracking and engagement algorithms, and systems engineers to integrate these various components into a cohesive defensive shield. If you are excited by a fast-paced environment where the threat and the technology are constantly changing, the world of counter-UAS is one of the most dynamic and in-demand fields in the entire defense industry.

The New Physics: Hypersonics and Directed Energy

Two disruptive technologies are forcing a fundamental rethinking of defense engineering, creating entirely new categories of platforms and effects that challenge the traditional boundaries of the domains. These are not incremental improvements; they are step-changes in capability that are governed by a different and far more extreme set of physical laws. They are incredibly difficult, incredibly expensive, and represent the absolute cutting edge of modern science and engineering.

Hypersonics: Vehicles that can travel at over five times the speed of sound (Mach 5), or more than one mile per second, in the upper atmosphere are a game-changing technology. They are incredibly difficult to detect and defend against due to their sheer speed and their ability to maneuver. This is not just a faster jet; it is a different kind of flight, operating in a flow regime where the air itself can break down into a plasma.

Directed Energy (DE): High-energy lasers and high-power microwaves are moving from science fiction to reality. These systems offer the promise of near-instantaneous, speed-of-light engagement with a virtually unlimited magazine (as long as you have electrical power). Instead of launching a physical interceptor, you are directing pure energy at a target to disable or destroy it.

The Brutal World of Hypersonics

The primary challenge of hypersonic flight is extreme thermal management. At Mach 5 and above, the friction of the air molecules creates temperatures on the leading edges of the vehicle that can exceed 2,000°C (3,600°F), hot enough to melt steel. This requires the invention and application of entirely new classes of advanced materials that can withstand these searing temperatures without losing their structural integrity.

The second great challenge is propulsion. A traditional jet engine cannot function at these speeds. The solution is the scramjet (supersonic combustion ramjet), a deceptively simple-looking engine with no moving parts. The scramjet uses the vehicle's incredible forward speed to compress the incoming air for combustion. The engineering challenge is immense: you must manage to inject fuel, mix it, and have a stable combustion event in a flow of air that is moving through the engine at supersonic speeds, a process that has been compared to "lighting a match in a hurricane."

This is the realm of the deep technical expert, particularly for mechanical and aerospace engineers. If you have a true passion for thermodynamics, heat transfer, computational fluid dynamics (CFD), and materials science, this is where you will find your frontier.

The Science of Directed Energy

Directed Energy weapons, primarily high-energy lasers (HEL), are a different kind of revolution. The core challenge is power and thermal management on a massive scale. To generate a laser beam powerful enough to burn through the skin of a drone or a missile, the system requires an immense amount of electrical power, often hundreds of kilowatts. Generating this power on a mobile platform like a truck or a ship is a major engineering problem.

Furthermore, modern solid-state lasers are not perfectly efficient. A significant portion of that electrical energy is converted not into light, but into waste heat. A 300-kilowatt laser that is 30% efficient is also a 700-kilowatt heater. Managing this waste heat, dissipating it so the laser itself doesn't overheat and destroy itself, is the primary limiting factor in the technology today. This requires incredibly sophisticated thermal management systems.

The final challenge is beam control. Hitting a small, fast-moving target miles away with a beam of light that is only a few inches wide requires a system of optics, mirrors, and tracking algorithms of almost unimaginable precision. The system must compensate for the turbulence in the atmosphere, which can distort the laser beam and cause it to lose its focus and its lethality.

This field is requires intimate understanding of how to communicate and operate on an interdisciplinary plane. It requires electrical engineers for the power electronics, mechanical engineers for the thermal management systems, optical engineers for the beam directors, and software engineers for the tracking and control algorithms.

The Strategic Implication

These technologies are not just engineering curiosities; they are strategically vital. They represent a fundamental shift in the offense-defense balance. Hypersonic weapons are difficult to stop; directed energy weapons are a potential solution. The nation that masters these technologies first will have a decisive advantage for decades to come.

Pursuing a career in these fields means you are stepping into the heart of the next great technological race. It is a world of deep R&D, of patient, foundational science, and of high-stakes competition. Pursuing a master's degree or Ph.D. in one of the core disciplines that underpin these technologies like plasma physics, advanced materials, or power electronics and aligning with the companies and national labs that are leading this research will place you at the very center of this competition. This is the path for the future Technical Fellow.

From Left of Boom to the Finishing Blow

The final and most strategic megatrend is not about a single technology, but about the philosophy of *when* technology is used. For decades, the primary focus of defense engineering was on winning a conflict after the first shot was fired, known in the industry as "right of boom." "Boom" signifies the moment a weapon detonates. While this remains critical, the new strategic focus is expanding across the entire timeline of conflict. This creates two new and powerful drivers for engineering innovation: operating "left of boom" to prevent wars from starting and seeking a decisive 'finishing blow' to end them quickly.

Left of Boom: The Engineering of Deterrence and Disruption

The most desirable victory is the one that is achieved without a fight. A huge and growing portion of the defense and intelligence world is dedicated to creating systems that operate "left of boom," before the shooting starts. The goal is deterrence and disruption. This is the world of intelligence services and of strategic, non-kinetic effects. It involves building the incredible reconnaissance satellites that can detect a potential adversary's preparations, providing leaders with the warning they need to de-escalate a crisis.

Left of Boom also encompasses the world of cyber and information operations—tools used to disrupt an adversary's ability to organize, to sow doubt, and to prevent them from ever reaching the point where they feel confident in launching an attack. This is a highly abstract and deeply intellectual arena, where the goal is to win the "war before the war."

This is the heartland of the intelligence community. It is a world that demands deep analytical skills, creativity, and the ability to think like an adversary. Computer scientists, data scientists, and electrical engineers specializing in signals intelligence are the primary players here. A career in this space means working on the nation's most sensitive and impactful programs, where the measure of success is a conflict that never happens.

Right of Boom: The Search for the Finishing Blow

When deterrence fails and a conflict begins ("boom"), the new strategic focus is on preventing long, drawn-out wars of attrition. The goal is to achieve a "finishing blow;" to deliver a decisive, overwhelming effect that can end the conflict quickly and on favorable terms. This requires systems that can create a level of shock and disruption from which an adversary cannot recover.

This is the world of hypersonic weapons, which can strike strategic targets anywhere in the world in a matter of minutes. It is the world of advanced electronic warfare systems that can dismantle an enemy's entire command and control network in the opening hours of a conflict. It is also the world of AI-driven, networked munitions, where a swarm of smart weapons can collaboratively overwhelm a sophisticated air defense system.

The engineering challenge here is immense. It requires a deep focus on speed, on power, and on networking that allows these capabilities to be brought to bear in a coordinated, overwhelming fashion.

You are now in the realm of the high-performance systems designer; a world for the aerospace and mechanical engineers working on hypersonics, the RF and software engineers creating the next generation of electronic attack systems, and the AI specialists designing the brains for collaborative munitions. A career here means working on the very tip of the spear, creating the technologies that will provide a decisive, conflict-ending advantage.

Your First Strategic Action

These five megatrends—the primacy of software, the rise of the digital twin, the shift to networked systems, the dawn of new physics, and the expansion of the strategic timeline—are the powerful currents that will shape your entire career. Understanding them is your first act of strategic foresight. It is the intelligence you need to look beyond the immediate requirements of your first job and to start making the smart, long-term investments in yourself that will ensure your relevance and success for decades to come. These trends are not abstract futures; they are the present-day reality being built in laboratories and design reviews right now. In the next chapter, we will take this strategic intelligence and turn it into an actionable plan, showing you how to build the specific, modern skillsets you will need to become an indispensable Engineer of 2040.

Chapter 13

The Modern Skillset

In the last chapter, we mapped the technological and strategic megatrends that are fundamentally reshaping the defense landscape. That was your intelligence briefing. This chapter will outline your action plan for applying this knowledge in the following chapters.

Understanding these trends is the first step; the next and most critical is to translate that foresight into a deliberate strategy for your own skill development. A trend is not something that happens *to you*; it is a wave that a strategic engineer learns to ride. Building a resilient, successful long-term career is about making a series of smart, informed investments in yourself that will pay dividends no matter how the future evolves.

The following is a guide to the four modern skillsets, the "unwritten curriculum," that will separate the successful engineers from the obsolete ones in the coming decades. These are the practical abilities you must consciously cultivate to become an indispensable Engineer of 2040. This chapter will serve as context for the remainder of this book, helping to build habits and practical abilities that will make you an indispensable engineer.

The T-Shaped Engineer

The primacy of software has shattered the traditional, rigid walls between engineering disciplines. As previously discussed, an F-35 is not a mechanical system with a computer; it is a flying supercomputer that happens to have a jet engine. This new reality demands engineers who are T-shaped; that is, an engineer possessing a deep, world-class expertise in their core discipline (the vertical bar of the 'T'), but also a broad, functional fluency in the adjacent disciplines they interact with (the horizontal bar).

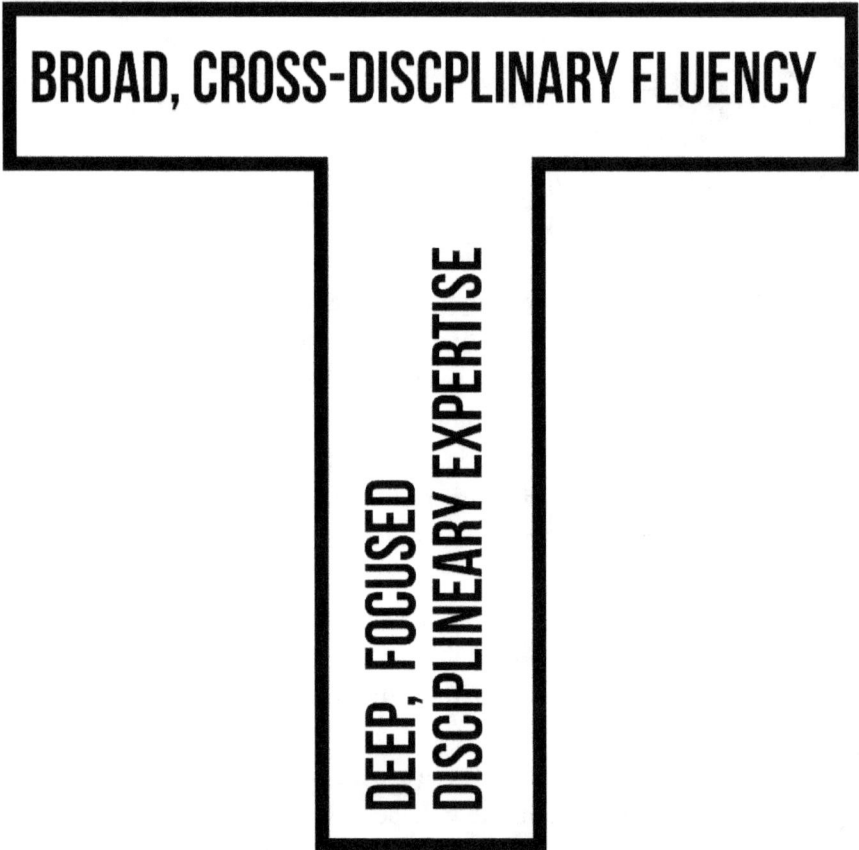

BROAD, CROSS-DISCPLINARY FLUENCY

DEEP, FOCUSED DISCIPLINEARY EXPERTISE

Figure 4: The T-Shaped Engineer. The modern engineer combines deep, vertical expertise in their primary discipline with a broad, horizontal fluency in the adjacent skills and domains needed for effective systems integration.

Today, that horizontal bar is overwhelmingly defined by software, data, and computation. An engineer who cannot "speak the language" of the other disciplines on their team is a bottleneck. An engineer who can is an invaluable integrator and a future leader.

For the mechanical & aerospace engineers, you must learn to code. Your immediate goal should be to achieve proficiency in Python. This is not about becoming a software engineer; it is about learning the language of data analysis and automation. An engineer who can write a script to parse gigabytes of test data from a vibration test or automate a series of FEA runs is ten times more effective, and ten times more valuable, than one who is trapped in the manual world of spreadsheets. Start with the scientific computing libraries: NumPy, Pandas, and Matplotlib.

For the electrical & computer engineer, you must understand the software that brings your hardware to life. The circuit board you design is a vessel for code. Your value skyrockets when you can have an intelligent conversation with the software team. Actively seek out courses in Embedded Systems, Real-Time Operating Systems (RTOS), or Computer Architecture. The EE who understands the timing constraints and memory limitations of the software that will run on their board designs better, more efficient hardware.

For the computer scientist, you must embrace physics. Your code will not live in a climate-controlled server farm; it will live on a piece of hardware that is hurtling through space, vibrating on a tank, or pulling 9 Gs. Your algorithms will be controlling physical things. Actively seek out an elective in Control Systems, Dynamics, or even basic circuit analysis. The CS graduate who understands the physical system they are controlling writes safer, more robust, and more effective code.

The Digital Engineer

The Digital Twin revolution is transforming engineering from a document-based process to a model-based one. The digital model *is* the authoritative source of truth. Your fluency with your respective discipline's core digital tools is not just a line item on a resume; it is the primary measure of your day-to-day professional competence and efficiency.

Your goal is to become a "power user." Do not be content with just knowing the basics of how to model a part in SOLIDWORKS or Creo. Go deeper. Learn the advanced surfacing tools. Understand how to properly use configurations and design tables. Successful engineers understand not just how to use the tool, but how to use it to express design intent, and by doing so, creating models that are robust, easy to update, and useful for others. The most direct way to prove this competence is to pursue the industry-recognized certifications offered by the software vendors. This clearly shows a hiring manager that you are a serious and dedicated user. The engineer who understands how their design connects to the wider world of analysis, manufacturing, and sustainment, sometimes referred to as the "digital thread," is an invaluable asset in the modern enterprise.

Designing for "X" will greatly streamline your design process and allow for greater success, earlier on. On your projects, practice Design for Manufacturability (DFM). Have a conversation with the university machinist. Ask them, "How would you actually make this part? Is this internal corner too sharp for your tools?" In practice, this will reduce turnaround time and redesign effort of any products that you own. More importantly, this shows a level of practical wisdom that interviewers are desperate to find.

Learn the "metagame" of your design tools. A CAD model does not exist in a vacuum; it is part of a much larger digital ecosystem. Go beyond your specific design tool and research the Product Data Management (PDM) and Product Lifecycle Management (PLM) software that are the backbones of large engineering efforts. Software like Siemens Teamcenter, Dassault Systèmes ENOVIA, or, more commonly, PTC Windchill are the digital libraries that manage all engineering data. Understanding that these systems are what enforce the strict configuration management and version control on a program of thousands of engineers shows that you are thinking at an enterprise level. You don't need to be an expert but being able to speak intelligently about the difference between a simple "file save" and a formal "check-in/check-out" process will instantly signal a higher level of professional maturity to a hiring manager.

The Systems Thinker

The rise of networked, autonomous, and collaborative systems represents a fundamental philosophical shift. In the 20th century, the goal was to optimize the performance of a single platform. In the 21st, the goal is to optimize the performance of the entire network. The value of your work will increasingly be measured not just by your component's individual performance, but by its ability to reliably and securely communicate and integrate with the larger system of systems.

This requires you to learn the language of interfaces. Every component you design is a node on a network. From day one, practice asking the critical systems-level questions: How does my part connect physically? How does it get power? How is it cooled? How does data get into it, and how does data get out of it? In a world of contested networks, every engineer must have a foundational understanding of cybersecurity. Is the data link for my new sensor encrypted? Is it vulnerable to jamming? Answering these questions is no longer just the software team's job. Cultivating this "systems thinking" mindset, where you are constantly thinking about the connections *between* the boxes and not just what's inside your own, will set you on the path to a leadership role.

Practice thinking in interfaces. On your capstone project, volunteer to create the Interface Control Document (ICD). This is a powerful demonstration of systems thinking. What are the specific connectors (mechanical)? What are the voltage and data lines (electrical)? What is the data format (software)? By using this skill, you will not only show your ability to decipher complex systems, but you are also making a case for yourself to be a future program leader.

Embrace block diagrams. The block diagram is how the systems thinker communicates. It is the single most powerful tool for cutting through complexity. Get in the habit of sketching one at the start of any problem, whether it's on a whiteboard or in your notebook. A good block diagram forces you to abstract away the details and focus on the system architecture: the major components (the blocks), the interfaces between them (the lines), and the flow of information or energy. It is the first and most important step in deconstructing a complex problem into a manageable set of smaller, solvable ones.

Learn foundational cybersecurity. In a world of networked and collaborative systems, cybersecurity is not an optional extra; it is a fundamental design requirement for all disciplines. Understanding the basics is no longer a niche skill for security specialists; it is a core competency for the modern engineer. This doesn't mean you need to become a hacker. It means you must be able to have an intelligent conversation about the "CIA Triad" (Confidentiality, Integrity, and Availability) of your system. It means understanding core concepts like encryption (protecting data at rest and in transit) and authentication (ensuring you are talking to who you think you are talking to). In your design reviews, you will be expected to answer the question: "How have you made this system secure?" This knowledge is the first step to giving a credible answer.

Mastering the Business of Engineering

In the academic world, you are trained to find the single, technically optimal solution to a well-defined problem. It is a search for perfection. In the professional world, engineering is the art of delivering the best possible solution within an unforgiving set of constraints. The two most fundamental and universal of those constraints are time (schedule) and money (budget).

An engineer who can only think in terms of technical elegance is an incomplete engineer. The true value of a senior engineer or a chief engineer is not just their technical brilliance, but their ability to make difficult, data-driven trade-offs between performance, cost, and schedule to deliver a solution that meets the customer's needs *on time and on budget*.

A hiring manager knows this. When they interview a student, they are not just looking for technical skill; they are looking for signs of professional maturity. Demonstrating that you have already practiced thinking in these terms and that you understand that a project has a budget, that an engineer's time is the most valuable resource, and that every decision is a trade-off is a massive and rare differentiator. It proves that you are not just a "book-smart" student; you are already thinking like a professional. It signals that you are the candidate who will be a quick study, a low-risk hire, and a future leader.

Create and manage a bill of materials (BOM). Create a detailed spreadsheet listing every single component, part number, supplier, and cost. This meticulous act forces you to engage with real-world budget constraints and the supply chain. More importantly, it becomes an incredibly powerful interview story. You will be able to tell a manager, "On my capstone project, I was responsible for the BOM. I managed over 50 line items and brought the project in 5% under our $1,500 budget." This is an answer that proves you already have business acumen.

Track your "labor" hours. Get in the habit of logging the hours you and your teammates spend on your capstone project. This gives you another quantifiable metric to communicate the scale of your effort.

Document your "make vs. buy" decisions. Engineering is the art of trade-offs and optimization. Actively look for opportunities to make a strategic make vs. buy trade-off and document the cost and schedule implications. Being able to explain this in an interview proves you are a systems-level, resource-conscious thinker who can make smart decisions to maximize the overall performance of a project within its given constraints.

These four skills, being T-shaped, mastering the digital thread, thinking in systems, and being a mission-focused pragmatist, are the hallmarks of the modern engineer. They are not discrete courses, but rather a mindset, an "unwritten curriculum" for a successful career. By consciously seeking out the projects, electives, and self-study opportunities to develop these abilities, you are not just preparing for your first job; you are building the intellectual foundation for your entire professional life. You are architecting a future-proof career.

Part 4

Forging Your Credentials (The College Years)

Chapter 14

The Strategic Coursework

You now have a detailed map of the defense industry. You understand its culture, its domains, and the technological horizon. The journey from being an interested student to a sought-after candidate begins now, and your primary battlefield for the next four years is your university. This is where you will forge the raw materials of your career. Every choice you make, from the electives you select to the projects you pour your weekends into, tells a story about the kind of engineer you are becoming. It is a story that recruiters are trained to read, and this chapter is your guide to writing a masterpiece.

Your core engineering curriculum is the price of admission. It is a grueling but requisite gauntlet of calculus, physics, statics, and thermodynamics by design. It proves you have the intellectual power to solve complex problems. But it's just the starting point. In a competitive field, a high GPA in your required courses makes you qualified, but it does not make you special. Your true differentiation lies in the choices you make on the margins. A defense contractor isn't just hiring a "mechanical engineer;" they are looking for a future Thermal Analyst, a future Robotics Engineer, or a future Energetics Specialist. Your academic and extracurricular choices are your first and best chance to signal which of those paths you are preparing for.

However, not every university offers a vast catalog of specialized defense-related electives. This chapter is a practical, strategic guide for students at *any* institution. We will focus first on mastering the foundational courses that are universally available and highly valued, turning your required learning into a deep, intuitive understanding. Then, we will cover how to add high-impact electives to build a specialty. Finally, and most importantly, we will explore how to signal your passion and expertise through action, from formal research to the invaluable skill of getting your hands dirty. This chapter is your guide to moving beyond simply *getting* an education and toward strategically *architecting* one.

Phase 1: Master the Foundations

Before you can specialize, you must first build an unshakeable foundation of first principles and practical skills. The following are the core courses and competencies that form the bedrock of a defense engineering career. These are the areas where you must do more than just get a good grade; you must achieve a deep, intuitive understanding and a high level of practical proficiency. Think of this as forging your core intellectual weapons, the tools you will rely on for the rest of your career.

For All Disciplines: The Universal Languages

There are a few subjects and skills that are so fundamental to the way the defense industry operates that they transcend individual engineering disciplines. These are the courses that teach you the language of physics. Mastering them isn't about passing exams; it's about developing an intuition for how the world works. This intuition is the true mark of a great engineer. Mastering these will allow you to communicate and collaborate effectively in the interdisciplinary world of large-scale programs.

CAD and Schematics: This is the most fundamental, practical skill you will learn. The language of engineering is not English; it is the drawing. Whether it's a 3D Computer-Aided Design model of a mechanical part or an electrical schematic of a circuit board, the digital model and the formal drawing are the contractual representation of the design. Your first six months on the job will likely be spent almost entirely inside a CAD tool. A senior engineer will give you a "red-lined" drawing, a marked-up printout of an existing design, and your task will be to incorporate those changes into the official 3D model and create a new, revised drawing that complies with all company and ASME standards.

Your ability to do this efficiently, accurately, and without needing your hand held is your first and most important test. Can you navigate a complex, thousand-part assembly to find the one component you need to modify? Do you understand the principles of design intent and how to model a part so that it can be easily updated later? Do you know how to create a clean, unambiguous drawing with proper GD&T callouts? Becoming a power user in a tool like SOLIDWORKS or Altium Designer is not just a resume builder; it is a direct investment in your day-one productivity and is the fastest way to earn the trust of your new team.

Control Systems: Nearly every defense platform is a complex control system. While you may never be the lead GNC engineer, you will absolutely be required to understand how your component fits into the larger dynamic system. In your interview, you might be asked, "Explain what a PID controller is and give me an example of where you would use one." Your ability to clearly explain Proportional, Integral, and Derivative gains and how they affect a system's response (e.g., controlling the temperature of an oven or the position of a robotic arm) is a fundamental test of your understanding.

On your first project, you may be asked to help model a small part of a larger system in MATLAB/Simulink. For example, you might be tasked with modeling the performance of a single electro-mechanical actuator in a missile's fin control system. You will be given the data sheet with the motor's torque and speed curves, and your job will be to create a Simulink block that accurately represents its real-world behavior. Your deep understanding of feedback, stability, and system dynamics from your coursework will be the foundation for this work. This is one of the most common entry-level tasks for engineers in the GNC (Guidance, Navigation, and Control) world.

Project-Based Design-Build-Test Courses

Your senior capstone design project is paramount. In your interview, you will be asked, in detail, about this project. They will not ask about your grade. The hiring manager will say, "Walk me through your senior design project. What was the goal? What part did *you* specifically own? What was the hardest technical problem you faced? What went wrong, and how did you fix it? What would you do differently if you did it again?"

Whether it be the late nights spent troubleshooting a noisy sensor on your circuit, the frustration of an FEA model that wouldn't converge, or the debate with your teammates over a design trade between weight and strength, the stories you generate here are the most valuable currency you have. They prove you can handle the messy, frustrating, and ultimately rewarding realities of engineering. A student who can tell a detailed story about debugging a problem is infinitely more hireable than a student who can only talk about their GPA.

For Mechanical & Aerospace Engineers

Heat Transfer: You will not be asked to derive the heat equation in an interview. You will be asked a practical question like, "I have a sealed electronics box with a 100-watt processor inside. What are the first three things you would consider to cool it?" Your ability to immediately start whiteboarding the problem and discussing the trade-offs between different solutions like passive conduction to the chassis, adding a heat sink and a fan for forced convection, or using high-emissivity coatings for radiation will prove your competence. Many junior ME tasks involve exactly this: performing a "first-order" thermal analysis of a circuit card or power supply to see if it's a problem that needs a deeper look.

Mechanical Vibrations: A senior engineer will not ask you to solve a complex differential equation on the board. They will hand you a Frequency Response Function (FRF) plot from a vibration test and ask, "What is this telling us?" Your ability to look at that plot and immediately identify the primary natural frequencies (the sharp peaks), to understand what a "random vibe" profile represents, and to suggest practical solutions like adding damping material or stiffening a bracket to shift a resonance, is a critical, real-world skill that shows you can interpret test data.

Fluid Mechanics / Aerodynamics: This knowledge is the price of admission for a career in the Air, Sea, or Space domains. Your first task might be to run a CFD analysis on a small, modified component, not the entire aircraft. For example, a senior engineer might ask you to analyze the effect of adding a new sensor pod to an existing pylon. Your understanding of boundary layers and compressible flow will be essential to setting up the model correctly, ensuring you have the right mesh density, and being able to interpret the results to see if the new pod creates any unexpected aerodynamic interference.

For Electrical & Computer Engineers

Electromagnetics: You won't be asked to recite Maxwell's equations. In an interview, you'll get a scenario: "You have a high-speed digital signal trace next to a sensitive analog trace on a PCB. What are you worried about, and what would you do to fix it?" Your answer, which should immediately bring up the concept of crosstalk (a core E&M problem), and your suggestions, like adding a ground trace in between or ensuring proper layer stack-up, will demonstrate your practical understanding.

Signals & Systems / Digital Signal Processing (DSP): Your first task will not be to invent a new filter. It will be to implement a known filter (like a Butterworth or a Chebyshev) in MATLAB, run a test signal with known noise characteristics through it, and verify that the output matches the theoretical performance. This is a fundamental "sanity check" task that every junior DSP engineer does. A deep, practical understanding of the Fourier Transform and its practical applications is necessary.

Embedded Systems: You will not be asked to design a new processor. You will be given a microcontroller (like an STM32), a technical data sheet that is 1,200 pages long, and a simple sensor. Your first task will be to write the C code to initialize the processor's I2C communication peripheral and read a value from the sensor. This simple-sounding task is a fundamental test of your ability to read dense technical documentation, configure registers, and write code that interacts with real hardware at the lowest level.

For Chemical Engineers

Thermodynamics & Transport Phenomena: While all engineers take thermodynamics, for a ChemE, this is home turf. A deep mastery of heat, mass, and momentum transfer is foundational for a career in advanced materials, thermal management, or propulsion. In an interview, you might be asked to explain the best way to manage the thermal decomposition of a specific material, a question that directly tests your intuitive understanding of these principles.

Kinetics & Reactor Design: This knowledge is the bedrock of energetics. Your first project might not be designing a new rocket motor, but helping a senior engineer analyze the stability and aging properties of a specific solid rocket propellant formulation. This involves using your understanding of reaction kinetics to model how the material's properties will change over a 20-year lifespan in storage, a task with critical safety implications.

Materials Science: A deep understanding of polymer science and metallurgy is crucial. You may be tasked with researching and recommending a specific epoxy or adhesive to bond a composite structure on a satellite. This is not a simple Google search. It requires you to dig into vendor data sheets, understand the effects of outgassing in a vacuum, degradation under UV exposure, and present a formal trade study to a senior engineer on why your chosen adhesive is the best option.

Phase 2: Add High-Impact Electives

Once you have mastered the foundations, you can use your limited elective slots to build a "spike" of specialized knowledge. These are the courses that, when a recruiter scans your transcript, will make them stop and take notice. They signal a level of interest and initiative that goes beyond the standard curriculum and show that you are already thinking like a professional.

Introduction to Systems Engineering: If your university offers a systems engineering course, take it. This is the single best signal you can send. It demonstrates that you understand the holistic, requirements-driven design philosophy that governs the entire industry. It teaches you the language of the V-model, of requirements traceability, and of interface management, putting you leagues ahead of your peers who are only thinking about their own specific component. In an interview, when you can talk about how your senior design project had a "System Requirements Review (SRR)," you are speaking the language of the industry.

Introduction to Finite Element Analysis (FEA): An FEA course demonstrates your ability to use the modern analysis tools of the trade. It proves you can move beyond simple textbook equations and into the world of complex, computer-aided engineering. It teaches you the critical art of engineering approximation: how to simplify a complex real-world problem into a model that is solvable but still accurate enough to be useful. Knowing the fundamentals of meshing, boundary conditions, and, most importantly, how to tell when your results are wrong, is a highly sought-after practical skill.

Manufacturing Processes / GD&T: For mechanical engineers, this is a course that proves you design parts that can actually be made. A deep understanding of Geometric Dimensioning and Tolerancing (GD&T) is the formal, authoritative language of professional mechanical design. Being able to have an intelligent conversation with a machinist about how your part will be fixtured, what the datums are, and whether a specific tolerance is achievable is a superpower that is surprisingly rare and highly valued. It shows that you are not just a "CAD jockey," but a true designer.

Introduction to Robotics / Mechatronics: This is the entry point into the world of autonomous systems. These courses are inherently inter-disciplinary, combining mechanical design, electronic sensors, and soft-ware control. They are a perfect way to demonstrate your ability to think at a systems level and are directly applicable to one of the fastest-growing and most exciting areas in the entire defense industry. When you can talk about how you integrated a motor, a sensor, and a microcontroller to achieve a goal, you are demonstrating the core skill of a systems engineer.

Phase 3: Signal Your Interest Through Action

What if you're passionate about hypersonics, but your university doesn't have a dedicated course? This is the reality for most students. You can still become a top candidate. In fact, proving your passion through in-dependent action can be even more impressive than simply taking a class. Here, you move from being a student to being a self-directed engineer.

Targeted Undergraduate Research: This is the most powerful method for gaining deep, specialized knowledge and a powerful mentor-ship relationship. Find a professor in your department (or even an adjacent one like materials science or physics) whose research aligns with your in-terests. Do not just send a generic email. Read their published papers.

Understand what they are working on. Go to their office hours and ask intelligent, informed questions about their research. Then, and only then, express your interest in contributing. Offer to help in their lab running basic tests or organizing data, even if it starts as a voluntary position. This shows incredible initiative and can often lead to a paid position, your name on a published paper, or a powerful, detailed letter of recommendation that will open doors.

A Strategically Chosen Personal Project: Use your personal or capstone project to build your own curriculum and become an expert in something you're passionate about. This is your chance to go deep.

Want to work on UAVs? Build one. Don't just assemble a kit. Design a custom airframe in SOLIDWORKS, perform a basic CFD analysis to optimize the wing shape, 3D print it, and use an open-source flight controller like Ardupilot to learn the basics of flight control logic and sensor integration. Document your entire process with a detailed report, photos, and video.

Interested in robotics? Build an autonomous rover that uses LIDAR for obstacle avoidance and computer vision to navigate. This is a great way to teach yourself the fundamentals of the Robot Operating System (ROS), a critical and highly sought-after skill in the world of autonomy.

Leverage Student Design Competitions: This is one of the single best experiences you can have in college. Joining a student design competition team is a fantastic way to gain deep, practical, hands-on experience in a specific domain. Teams like Formula SAE (for automotive engineering), Design/Build/Fly (for aerospace), Baja SAE (for off-road vehicles), or a university rocketry club are effectively mini-programs where you will deal with real budgets, deadlines, and teamwork challenges. Your role as the "lead for the aerodynamics sub-team" or the "powertrain lead" is a powerful resume entry that allows you to tell compelling, STAR-method stories that are not theoretical, but grounded in the real-world constraints of budgets, deadlines, and unexpected failures.

Online Learning & Self-Study: While an online certificate from Coursera or edX doesn't replace a university course, it does show initiative and a passion for learning. The key is to not just collect certificates. If you're interested in AI/ML, completing a well-regarded online course from a top university and then applying those skills to a small, personal project is a great way to demonstrate your interest and build a practical skill set that you can actually talk about in an interview.

Phase 4: Get Your Hands Dirty

There are two types of engineers in the world: those who can get good grades and those who can *do*. The engineers who can do both are the ones who become legends. In an industry that builds real, tangible, high-consequence hardware, the value of practical, hands-on experience cannot be overstated. It is often the single greatest differentiator between a good candidate and a great one.

Hiring managers know that a student who has spent their weekends rebuilding an engine in their garage, troubleshooting a complex electrical problem on their own 3D printer, or building their own electronic circuits has learned a set of invaluable lessons that are not taught in any classroom. They have learned how to diagnose a problem when there's no answer in the back of the book. They have learned the frustration of a stripped bolt or a cold solder joint. They have developed an intuitive "feel" for how real-world hardware actually works and how it fails. This practical knowledge is a superpower.

The Power of a Tangible Hobby

This is about finding a passion that gets you away from the computer screen and into the physical world. It doesn't have to be directly related to your career goal; it just has to involve making and fixing things.

Do you love cars? Get a project car. Working on cars is always a great source of hands-on experience. The process of taking apart an engine, understanding how the systems work together, and troubleshooting a weird electrical issue will teach you more about systems integration and practical problem-solving than a dozen textbooks.

Are you an EE or CompE? Get into hobby electronics. Learn to solder. Build your own circuits from scratch using an Arduino or Raspberry Pi. Design and etch your own printed circuit boards. This deep, practical knowledge of how electrons actually flow is what separates the great EEs from the purely theoretical ones.

To gain experience with machine and fabrication, join your university's machine shop or a local makerspace. Learn to use a manual mill and a lathe. Learn to weld. The mechanical engineer who understands the realities of fabrication, who knows what is easy and what is difficult to make, will always design better, cheaper, and more effective parts.

How to Showcase This on a Resume

You might think a "hobby" doesn't belong on a professional resume, but you'd be wrong. You can and should include it, but it must be framed professionally. Don't just list "Cars" as a hobby. Create a small "Personal Projects & Skills" section.

Example: "Automotive Restoration & Diagnosis: Actively involved in the complete mechanical and electrical restoration of a 1988 BMW E30, including engine diagnostics, suspension overhaul, and wiring harness repair."

Example: "Hobby Electronics & PCB Design: Proficient in soldering and circuit board rework. Designed and fabricated custom control boards for multiple Arduino-based robotics projects."

Why It Matters in the Interview

This hands-on experience is a goldmine of stories for your interview. When an interviewer asks, "Tell me about a time you solved a difficult technical problem," you don't have to rely on a homework assignment. You can tell the story of the three days you spent trying to diagnose a bizarre electrical short in your car, and the "aha" moment when you finally traced it to a corroded ground wire. This is a real story of grit, of persistence, and of practical problem-solving. It is authentic, it is memorable, and it is exactly what hiring managers want to hear.

Your transcript proves you have the potential to be an engineer. Your hands-on skills prove that you already are one. By mastering the fundamentals, taking advantage of high-impact electives when you can, and proactively demonstrating your passion through research, projects, and tangible skills, you can build a compelling profile that makes you a top-tier candidate, no matter where you went to school.

Chapter 15

Beyond the GPA

Your GPA and your coursework are your academic credentials. They prove you have the required technical knowledge and the discipline to succeed in a rigorous academic environment. But in the eyes of a recruiter, they are just table stakes. They are the price of admission, the minimum requirement to get your resume a first look. They tell a hiring manager what you *know*, but they do not tell them what you can *do*. As we have touched on in the previous chapter, your transcript is a record of your past performance in a structured, academic world; it is not a predictor of your future performance in the messy, unstructured, and often chaotic world of real engineering.

To answer the question of what you can *do*, you must go beyond the GPA. In a stack of a hundred resumes with similar grades from similar universities, the one that gets moved to the top of the pile is the one that can answer the question:" But what have you *built*?"

This is where a personal project, or more importantly, a significant academic team project like your Senior Capstone, becomes your single most powerful asset. A project is tangible, undeniable proof of your skills. It transforms you from a "student with a 3.7 GPA" into "the student who designed and built the autonomous rover that took third place in the university competition." It gives you a story to tell, a focal point for your resume, and a wellspring of confident, detailed answers for the toughest interview questions. This chapter will deep dive into projects and act as your guide to choosing, executing, showcasing, and ultimately *weaponizing* your project work to give you a decisive edge.

Phase 1: The Strategy

Phase 1 focuses on the upfront thinking and planning that separates a simple hobby from a strategic career asset. In the world of engineering, what you choose to build reflects what you want to become. A project is not just a demonstration of skill; it can be viewed as a declaration of intent. A well-chosen project demonstrates not just technical aptitude, but also forethought, passion, and a clear sense of direction. It becomes the compelling centerpiece of your professional narrative, the tangible evidence that your interest in a specific field is more than just a line on a cover letter. This is your first opportunity to begin crafting the story of the kind of engineer you want to be, long before you ever step into an interview room.

Choosing Your Project (For Those With a Choice)

This section is for the students who have a degree of freedom in selecting their projects, whether it's choosing from a list of capstone options or deciding what to build in your own time. This is a significant and often undervalued opportunity to align your academic work with your career ambitions. However, it's crucial to state with absolute clarity that if your curriculum assigns you a project, that is perfectly fine. Do not feel disadvantaged in any way. The most important thing is never *what* the project is, but the rigor, professionalism, and passion with which you execute it. A project to design a better can opener, when executed with meticulous analysis, brilliant documentation, and a deep understanding of the manufacturing process, can be far more impressive than a poorly executed drone project. Any project, when approached with excellence, can be the source of the powerful stories we will discuss.

For those with a choice, however, the selection of a significant project is one of your first major career decisions.

Align with Your Target Domain: Use the knowledge you've gained from the "Arenas of Innovation" section of this book to make a strategic choice. If you are passionate about the Air Domain, a project involving a 3D-printed RC plane, a drone, or a high-powered rocket is a powerful and direct signal of your interest. A recruiter for a naval systems company who sees an autonomous boat project on your resume immediately knows that you are not just applying for *a* job, but that you are a genuine member of *their tribe*. This alignment creates an instant, powerful connection and gives you a massive advantage in the interview as you can speak with genuine, demonstrated passion about their field.

Complexity is Good, Finishing is Better: It is incredibly tempting, especially for ambitious young engineers, to take on a massively complex project with a dozen cutting-edge features. Be wary of this temptation. A project that is 50% finished is a story without an ending. It leaves the interviewer with questions about your ability to manage scope and to deliver a final product. In contrast, it is far, far better to have a simple, 100% completed, and well-documented project than an incredibly ambitious, complex project that is only 50% finished. A finished project tells a complete story with a satisfying conclusion (the "R" in the STAR method). Start with a simple, achievable goal, like your "Minimum Viable Product," and then, only after you have achieved that goal, begin adding the more ambitious stretch features.

The "Borrowed" Project is a Valid Project: Not everyone has the time, resources, or workshop space for a massive personal project from scratch. That is perfectly fine. You can "borrow" a project. Your Senior Capstone Project is the most obvious and powerful example, and we will do a deep dive on this later. But you can also leverage a significant class project from one of your junior-level courses, taking it a step further than the assignment required. Or, as we discussed in the last chapter, you can join a student design competition like Formula SAE or Design/Build/Fly. Your role on that team, and the specific subsystem you own (like the "suspension IPT" or the "avionics bay"), becomes your project. These team-based projects are often even more valuable, as they provide a rich source of stories about teamwork, leadership, and conflict resolution.

A Brainstorming Guide to Projects by Domain

To help you align your project with your career ambitions, here is a list of tangible, exciting, and achievable project ideas, broken down by the five major domains. Use this not as a prescriptive list, but as a catalyst for your own creativity.

Air Domain Project Ideas:

The Classic: The Custom Drone. This is a fantastic way to demonstrate your understanding of aerodynamics, structures, and the all-important war on weight. Don't just assemble a kit. Design your own airframe in CAD, use FEA to optimize it for strength and low weight, and 3D print it. This proves you understand the full digital-to-physical workflow.

The Advanced: The Autonomous Drone. Take the classic project a step further. Create a simple, autonomous "follow me" drone. Using an open-source flight controller like a Pixhawk and a companion computer like a Raspberry Pi, implement a basic computer vision algorithm (*like a simple color-blob tracker*) using a library like OpenCV that allows the drone to visually track and follow a person or a brightly colored object.

The Rocketry Project: For those interested in propulsion and high-speed flight, building a high-powered model rocket is a perfect project. Go beyond the basic kit. Design it with a multi-stage deployment system for the parachute and, most importantly, include a telemetry payload with an Arduino and sensors that can transmit real-time altitude and acceleration data to a ground station you build.

Land Domain Project Ideas:

The Classic: The Rugged Rover. Build a rugged, wheeled or tracked robotic vehicle designed to navigate an outdoor obstacle course. Focus on the mechanical design of a robust chassis and a high-torque drivetrain that can handle rough terrain, grass, and inclines. This is a direct, tangible demonstration of skills relevant to combat vehicle design.

The Advanced: The Mapping Robot. Add a sensor suite (like LIDAR or a stereo camera) to your robotic vehicle and implement a basic SLAM (Simultaneous Localization and Mapping) algorithm. This is a powerful software and systems integration challenge that is at the heart of modern autonomous navigation.

The Mechanical Deep Dive: Design and build a miniature, working model of an advanced suspension system, like a double-wishbone for a wheeled vehicle or a Christie suspension for a tracked vehicle. Instrument it with sensors to measure and demonstrate its performance characteristics under different loads.

Sea Domain Project Ideas:

The Classic: The Autonomous Boat. Build a small, autonomous boat (an Unmanned Surface Vehicle - USV) that can navigate to a series of pre-programmed GPS waypoints in a local lake or pond. This is a great introduction to the challenges of marine vehicle control and waterproofing.

The Advanced: The "Sonobuoy." Create a "sonobuoy" style device: a waterproof, floating sensor package that you can deploy from a boat. Design it to measure water temperature and depth and transmit the data wirelessly via a LoRa (Long Range) radio module back to a base station you build. This is a great project for an aspiring naval systems engineer.

The Submersible: This is a major challenge, but incredibly impressive. Design and build a small, remotely operated underwater vehicle (ROV) with a camera feed. The primary engineering challenge to focus on is the immense difficulty of waterproofing the electronics, motors, and cable penetrations against water pressure.

Space Domain Project Ideas:

The Classic: The CubeSat Structure. Design and build a "CubeSat" frame. Using the official CubeSat design specification, create a 1U or 3U satellite frame in CAD and then fabricate it (either 3D printed or machined from aluminum). The goal is to create a structure that is both incredibly lightweight and incredibly strong and stiff, a core challenge of space engineering.

The Advanced: The Attitude Control System. Create a simple, working reaction wheel attitude control system for a mock satellite. Using a spinning flywheel, a motor, and a microcontroller with an IMU (Inertial Measurement Unit), write the code to demonstrate how a satellite can be precisely pointed by controlling the speed of the wheel. This is a fantastic and highly impressive control systems project.

The Ground Station: Build a motorized antenna tracking system that can automatically follow and receive signals from weather satellites (like the NOAA satellites) or the International Space Station. This is a fantastic RF and control systems project.

Cyber & EW Domain Project Ideas:

The Classic: The ADS-B Receiver. Use a low-cost Software Defined Radio (SDR) dongle and open-source software like GNU Radio to build a simple ADS-B receiver. This will allow you to receive and decode the unencrypted transponder signals from real aircraft flying overhead, displaying their location, altitude, and speed on a map. This would display a strong interest in signals intelligence.

The Advanced: The "Wardriving" Rig. Create a "wardriving" setup with a Raspberry Pi, a Wi-Fi adapter, and a GPS module. Write a Python script that logs the location and security status (WEP, WPA2, etc.) of all the Wi-Fi networks your device discovers as you move around. This is a practical demonstration of your understanding of network reconnaissance.

The Security Project: The Honeypot. Build a simple "honeypot," a deliberately vulnerable system on your network (like an old, un-patched Raspberry Pi), and use network monitoring tools like Wireshark to observe and analyze the traffic of any automated bots or attackers that try to probe it. This is a hands-on introduction to defensive cyber operations.

Phase 2: Executing A Project Like A Professional

You have a mission. You have chosen a project that is ambitious, achievable, and aligned with your career goals. Now comes the most important part: the execution. This is where you move from planning to doing, from strategy to engineering. How you execute your project is as important as what you build. A brilliant design that is poorly documented or a project that ignores real-world constraints is a missed opportunity.

This phase is a deep masterclass on *how to run your project like a professional*, a skill that will instantly differentiate you from your peers who are still thinking like students. We will cover the three pillars of professional execution: the art of documentation, the business of resource management, and a deep dive into your single most important project, your senior capstone.

Documentation As A Differentiator

This is the single most important and most often neglected part of any student project. Let's be blunt; if you do not document your work, it is just a hobby. If you meticulously document it, it becomes professional engineering experience. In the defense industry, the saying is, "The job isn't done until the paperwork is done." This is not a joke. Your documentation is the evidence that transforms your claims into credible facts. It is the proof that you can not only solve a problem, but that you can communicate that solution in a professional, organized, and permanent manner, a skill that is absolutely essential in a world of formal design reviews and configuration management.

Your goal is to create a "Project Portfolio" for your most significant project. This is a simple website (using a free tool like Google Sites or GitHub Pages) or a shared folder (like a Google Drive folder with a public link) that you can link to directly from your resume, right under the project's title. This is your first chance to live by one of the most critical unwritten rules of the defense industry: Process is King. In this world, an undocumented success is not a success at all. Your documentation transforms a personal achievement into professional evidence.

This portfolio is your evidence locker. It should be professional, well-organized, and contain these key elements:

A Clear Project Summary: This is the "abstract" of your project. In a short, well-written paragraph, explain the goal of the project, the problem you were trying to solve, and the final, quantifiable outcome.

High-Quality Photos and Videos: Humans are visual creatures. High-quality pictures of your finished product, and, if possible, of the building and testing process, are incredibly powerful. An even more powerful tool is a short, 30-60 second video of your project in action. A video of your drone flying, your robot navigating an obstacle, or your antenna tracking a satellite is a thousand times more compelling than a bullet point on a resume. It is undeniable proof of a working system.

A Clear Explanation of Your Contributions: On a team project, you must be explicit about your specific role. What were you the "owner" of? What was your specific subsystem? A simple, bulleted list detailing your responsibilities is perfect.

The Evidence (The Most Important Part): This is where you provide the proof of your engineering effort.

Final Report: Include a clean, well-formatted PDF of your final project report.

Key Artifacts: This is your chance to show off your best work. Include a few high-resolution screenshots of your most complex CAD models, your most detailed schematics, and your most impressive FEA plots or simulation results, each with a brief, explanatory caption.

Code Repository: If you wrote code for the project, you must include a link to your GitHub repository. This is non-negotiable for software, computer, and robotics engineers. A well-organized, well-commented GitHub, with a clear ReadMe file, is a powerful testament to your skills as a developer and your commitment to professional practices. For mechanical or electrical engineers, while a public code repository might not be your primary artifact, any scripts you wrote for data analysis (e.g., in MATLAB or Python) should also be cleaned up, commented, and included. This demonstrates the computational literacy we discussed in the last chapter.

Budgeting and Resource Management

There is a crucial distinction between a student and a professional engineer. A student is often asked to solve a technical problem with the primary goal of finding the most elegant or highest-performing solution. A professional engineer is always asked to find the best possible solution within a given set of constraints, and the two most fundamental constraints in the professional world are time and money. This section is about learning to think like a professional. By practicing these skills on your academic projects, you are demonstrating a level of professional maturity that is incredibly rare in a student, proving that you are a candidate who will spend their valuable time learning the company's complex technical challenges, not learning basic administrative tasks.

The Bill of Materials

The Bill of Materials (BOM) is one of the most fundamental and important documents in all of engineering. It is a detailed, line-by-line list of every single component that goes into your project. For a professional engineer, the BOM is the DNA of the design, the master document that connects the engineering world to the worlds of finance, procurement, and manufacturing.

How to Build It: Create a detailed spreadsheet for your project. The columns should include Item Number, Part Name, Description, Supplier, Supplier Part Number, Quantity, Cost Per Unit, and Total Cost. Be meticulous. Every single resistor, every bolt, every 3D-printed part should have its own line item. For the 3D-printed parts, you can estimate the material cost.

Why It's So Powerful: Creating and managing a BOM teaches you several critical professional skills. It forces you to be organized. It forces you to research and select real-world components from real-world suppliers like Digi-Key, McMaster-Carr, or Mouser. Most importantly, it gives you a powerful tool for your interview. Being able to say, "I managed a BOM with over 50-line items and a total budget of $1,500, and we came in 5% under budget," is an incredibly impressive, data-driven statement that proves you understand the financial side of a project.

The Make vs. Buy Decision

The "Make vs. Buy" decision is a classic engineering trade-off that you will face constantly in your professional career. The question is simple: is it better for the project to make a component from scratch (giving you complete design control but costing you time and manufacturing risk) or to buy a pre-made, commercial off-the-shelf (COTS) component (saving you time but limiting your design options)?

How to Practice It: On your project, actively look for opportunities to make a deliberate make vs. buy decision. For example, for your robot's chassis, you could have it professionally machined for $500, but your budget is tight. Instead, you could decide to "make" it in-house using the university's waterjet, which only costs $50 in raw materials. This decision frees up over $400 in your budget. You can then use that freed-up money to" buy" a much better, higher-performance LIDAR sensor than your original budget would have allowed.

Why It's So Powerful: This tells a compelling story about you thinking like a systems engineer. In an interview, you can say, "I made a strategic 'make vs. buy' trade-off on the chassis. By choosing to fabricate it in-house, I was able to save 80% of the budgeted cost for that component, which I then reallocated to purchase a more capable LIDAR sensor. This decision directly improved the overall performance of our robot's navigation system." This is exactly the kind of smart, budget-conscious, performance-focused thinking that hiring managers are desperate to find.

Tracking Your Labor Hours

In the professional world, the single most expensive and valuable resource on any project is not the materials or the software; it is the time of the engineers. Companies live and die by their ability to accurately estimate and track the number of "labor hours" required to complete a task. Demonstrating that you understand this fundamental business reality will immediately set you apart from your peers.

How to Do It: You don't need a complex system. Keep a simple log in your engineering notebook or a spreadsheet. At the end of each week, take ten minutes to jot down a rough estimate of how many hours you and your teammates spent on the project.

Why It's So Powerful: It gives you another incredible, quantifiable metric for your resume and your interviews. Even if it's an academic project with no "real" budget, being able to say, "Our four-person team invested over 400 manhours into this project over the course of the semester," is a powerful way to communicate the scale and the seriousness of the effort. It shows that you understand that an engineer's time is not free, and that you have a mindset that is already aligned with the realities of professional project management.

Your Capstone Project: A Professional Simulation

Your Senior Design / Capstone project is a gift. It is a semester-long, or even a year-long, built-in opportunity to get a fantastic, resume-defining project on the books. Do not coast through it. This is your chance to shine, to apply everything you have learned, and to create the centerpiece of your job application. It is the closest you will come to a real-world engineering program in your academic career.

The Capstone as a Program Simulation: Frame the capstone not as a class, but as a simulation of a real defense program. Your professor is your "Functional Manager," your team is your "IPT," and your sponsor (if you have one) is your "Customer."

A Guide to Team Dynamics: This is your chance to learn how to work with other engineers, and it will not always be easy. You will have to deal with "the slacker" who doesn't pull their weight, "the perfectionist" who never wants to finish a design, and the inevitable technical disagreements. Learning how to navigate these challenges, how to have a professional disagreement, and how to motivate your teammates is an invaluable skill. Stepping up as a leader in these situations is a powerful story for an interview.

How to Run Your First Design Review: Your project will have a PDR and a CDR. Treat them with the seriousness of a professional design review. Create a professional slide deck. Rehearse your presentation. Anticipate the tough questions your professors will ask. This is not just a class presentation; it is a professional rehearsal. Treat your PDR and CDR with the seriousness of a real design review. This is your chance to practice, in a low-stakes environment, the communication and justification skills that are absolutely critical in a defense career.

Phase 3: The Payoff

You have chosen your mission. You have executed it with the rigor and discipline of a professional. You have a finished, well-documented project that serves as evidence of your skills. Now comes the final and most important phase: the payoff. This is where you learn to "weaponize" your project, to transform it from a piece of engineering into a powerful and persuasive tool for your job hunt. All the effort you've put into executing and documenting your project is useless if you can't communicate its value effectively to a recruiter and a hiring manager.

This is not a time for humility. This is the time to confidently and professionally showcase the evidence of your capabilities. This phase is entirely focused on how to translate your hard work into a job offer. We will cover how to turn your experiences into compelling interview stories, the common mistakes that can sabotage your efforts, and the specific ways you can leverage each part of your project to answer the toughest questions.

How to Talk About Your Projects in an Interview

This is where all of your hard work pays off. Your project is not just a line on your resume; it is your goldmine of compelling stories for the behavioral interview. The single most common and predictable question you will be asked in any engineering interview is some version of, "So, walk me through your senior design project." This is not a casual question; it is an invitation. It is your chance to shine, to go beyond the bullet points, and to give the interviewer a case study of you being an effective, problem-solving engineer. A well-prepared answer to this question can set the tone for the entire interview.

Every significant project contains at least three powerful stories that you should prepare in advance, using the STAR method we will detail in the interview chapter. Do not try to come up with these on the fly. You should have them pre-scripted, practiced, and ready to deploy.

A technical challenge will be your primary story. It should detail the hardest technical problem you faced and the methodical, engineering-driven process you used to solve it. It should be rich with technical detail (the software you used, the analysis you performed) and end with a clear, quantifiable result.

If you were on a team, you must have a story about how you navigated a team dynamic. This could be a time you had to resolve a technical disagreement, a time you had to step up and lead a specific effort, or a time you helped a teammate who was struggling. This proves you have the professional skills to work effectively in a collaborative environment.

A story of failure and learning is the most powerful story of all.. Interviewers will often ask, "Tell me about a time something went wrong." They are not trying to trap you; they are testing your maturity and your ability to learn from your mistakes. You must have a story about a test that failed, a part that broke, or a simulation that gave you the wrong answer. The key is not the failure itself, but a clear, honest explanation of *what you learned from it* and how you would do things differently in the future.

Develop a "one-minute pitch" for your project. You must have a concise, powerful, one-minute summary of your primary project ready to go. This is your answer to that first, open-ended question. It should be a verbal abstract of your project, a high-level overview that is packed with compelling details.

> *Good Answer:* "My project was to build a robot. I was on the mechanical team. We designed the chassis in SOLIDWORKS, and it worked." (Factually correct, but boring and lacks impact.)

> *Gold Standard Answer:* "Our goal was to build an autonomous search-and-rescue rover for the university's annual competition, with a primary requirement of navigating an obstacle course and

identifying a target. I was the lead for the four-person mechanical sub-team, and I was personally responsible for the entire chassis and manipulator arm. The biggest technical challenge we faced was designing the chassis to be both lightweight and strong enough to survive a 3-foot drop test. To solve that, I used ANSYS to run an iterative FEA process that allowed us to reduce the frame weight by 15% while still achieving a 2.5 factor of safety. Ultimately, our team placed third in the competition, and the biggest lesson I learned was about the importance of design for manufacturability, which is a skill I'm excited to bring to a real-world program." *This represents a compelling narrative that showcases technical skills, problem-solving, leadership, and professional maturity.*

Common Pitfalls to Avoid

As you build, document, and talk about your projects, be aware of these common mistakes that can undermine your credibility. A single one of these errors can undo all of your hard work.

The "I" vs. "We" Problem: This is the most common mistake students make in an interview. When asked about a team project, they constantly say "we did this" and "we did that." While this shows you are a team player, it completely obscures your individual contribution and leaves the interviewer wondering what *you* actually accomplished. You must learn to take ownership of your work. Be specific.

Instead of: "We designed the circuit board."

Say: "I was responsible for the power regulation section of the circuit board, which I designed in Altium. My teammate, John, was responsible for the microcontroller section."

Forgetting to Document: This is the cardinal sin of student projects. A brilliant, complex project with no documentation is just a hobby. A simpler project with a professional, detailed portfolio is engineering experience. The engineer who shows up to an interview with a link on their resume to their project portfolio that is filled with reports, photos, and a link to their GitHub is in a completely different league from the one who just talks about what they did. It is the ultimate "show, don't tell."

Exaggerating Your Role: The temptation to overstate your contribution can be strong, especially if you were on a large team and feel insecure about your role. Do not give in to it. The engineering world is small, and an experienced interviewer can easily spot an exaggeration by asking a few detailed, probing follow-up questions. If you say you were the "lead," be prepared to answer specific questions about how you managed tasks, how you created the schedule, and how you resolved conflicts. Honesty and a clear, accurate description of your role are of the utmost importance. Losing your credibility is a fatal, unrecoverable error in an interview.

Weaponizing Your Project

This is the final step, where we connect all the dots. Every piece of work you have done in the execution phase can be "weaponized" as a direct, powerful answer to a specific interview question.

The Portfolio Link as a "Secret Weapon": The portfolio link on your resume is a "call to action" for the recruiter and the hiring manager. While other students just make claims, you are providing a link to the *evidence*. It is a sign of confidence and professionalism. In an interview, you can say, "As you can see in the final report in my online portfolio, we were able to..." This is an incredibly powerful way to substantiate your claims.

The BOM as a "Business Acumen" Story: When the interviewer asks, "Tell me about how you've handled a budget," most students will panic and say they have no experience. You will not. You will say, "On my capstone project, I was responsible for creating and managing the Bill of Materials. I tracked over 50 line items from three different suppliers to stay within our $1,500 budget, and we ultimately came in 5% under budget." This is a gold-standard answer that proves you already think like a professional.

Troubleshooting as Your "Grit" Story: The stories of your failures and how you fixed them are often more powerful than the stories of your successes. When an interviewer asks, "Tell me about a time you faced a challenge," you can tell the story of the three days you spent trying to diagnose a bizarre electrical short in your car, and the "aha" moment when you finally traced it to a corroded ground wire. This is a story of grit, of persistence, and of real-world problem-solving. It proves you are not just a "book smart" engineer, but someone who knows how to get their hands dirty and solve a problem.

From Student to Engineer

The journey through your undergraduate projects—from the first simple class assignment to the final, complex capstone—is the most important transformation of your academic career. It is the foundation where the abstract theory of your coursework is forged into the tangible, hard-won wisdom of a practicing engineer. This is the moment your identity changes. It is where you make the profound leap from being a student who learns about engineering to becoming an engineer who does engineering.

You now understand the deep, strategic "why" behind what a project demonstrates: the application of theory, the resilience of problem-solving, and the professional maturity of teamwork and communication. You have a guide for choosing a project, a brainstorming list of ideas, and a clear understanding of the critical, non-negotiable importance of documentation and resource management. You have a strategy for treating your capstone as your first job and a framework for transforming your project work into the compelling narratives that will dominate your interviews.

Your GPA is a snapshot of your ability to succeed in the structured world of academia. Your project portfolio is proof to your ability to succeed in the messy, unpredictable, and ultimately more rewarding world of real engineering. It is the proof that you are not just a student with a high grade, but a builder, a problem-solver, and a creator. In the chapters that follow, we will take this powerful, evidence-based foundation and use it to build the application arsenal that will launch your career. But always remember, the strength of that arsenal, the power of your resume, and the confidence in your interviews is forged right here. It is forged in the projects you choose, the rigor with which you execute them, and the professional stories you earn the right to tell.

Chapter 16

The Internship Gauntlet

Everything we have discussed so far, the strategic coursework, the hands-on projects, the late nights on a design team, has been in preparation for this single, critical goal: landing a high-quality internship in the defense and aerospace industry. This is not just another step; for many, it is the most important step of their entire undergraduate career, the moment when the theoretical world of academia gives way to the applied reality of professional engineering.

Let's be blunt: In the modern engineering landscape, an internship is no longer a "nice-to-have" or a simple resume booster. It is a mandatory prerequisite for a top-tier job. And while an internship with real experience in *any* technical field is valuable (proving you can work in a professional environment is always better than nothing), an internship *within* the defense and aerospace industry is a massive force multiplier. It is your trial run. It is the company's extended, three-month-long interview to see if you have the technical skills, the professional maturity, and the temperament to succeed in their culture. More importantly, it is your one and only chance as a student to get your foot in the door and begin the lengthy security clearance process, which is the single biggest hurdle to getting a full-time offer upon graduation.

Students with a relevant defense internship under their belt are a completely different class of candidate from those without. They are a known quantity. They have proven they can navigate a professional, classified environment, and they are already "in the system." Graduating with one, or preferably two, defense internships is the closest thing to a guaranteed full-time job offer you can get. This chapter is your playbook for the internship gauntlet; how to find the opportunities, how to land one, and how to perform so well that they can't afford to let you go.

The Hunt

The first phase of the gauntlet is the hunt. This is a proactive, intelligence-driven campaign that you will wage to find and identify the right opportunities. It is not a passive activity. Simply uploading a generic resume to a few online portals and hoping for the best is a strategy for failure. A successful hunt requires discipline, organization, and a deep understanding of the unique recruiting timeline of the defense industry. It is a marathon that begins long before you ever shake a recruiter's hand, and the work you put in here will directly determine the quality and quantity of the opportunities you receive.

Understanding the Calendar: Why the Rush?

The defense industry's recruiting cycle starts much, much earlier than you think. For a summer internship, the prime application season is September through November of the *previous year*. This is not an exaggeration. The primary reason for this is the security clearance process. It can take anywhere from six to eighteen months for the government to complete a background investigation. Therefore, for a company to have an intern with a clearance ready to start work in June, they must identify and start the process with that candidate in the early fall of the preceding year. Some intelligence agencies, like the CIA, are even more extreme, often opening their applications a full year or more in advance. You must align your personal schedule to this reality.

The Power of an Early Start

While it is more difficult to land a high-profile internship at a major Prime Contractor early in your college career, it is absolutely critical that you start trying from your freshman year. Every interaction is valuable practice. Think of it as a series of low-stakes trial runs: going to the career fair, talking to recruiters, and applying for positions. Even if you don't get them, the experience builds the confidence and polish that will make you a much stronger candidate in your all-important junior year, when the stakes are highest.

Getting your foot in the door early, however, is not impossible. It simply requires a more strategic approach.

Start with a focus on Federally Funded Research and Development Centers (FFRDCs) and university-affiliated labs. FFRDCs and University Affiliated Research Centers (UARCs), like Johns Hopkins APL, MIT Lincoln Laboratory, or the Aerospace Corporation, are often more open to hiring talented younger students for summer research positions. Their connection to the academic world makes them more comfortable with candidates who have strong theoretical knowledge but less project experience.

A story of failure and learning is the most powerful story of all too proud for your first role. Your first internship might be less about pure design and more about being a valuable apprentice supporting the core work of a team. You might be a "gopher," running tests for a senior engineer, organizing and plotting data, or helping a graduate student with their research. This is not a failure; it is a massive success. You are getting your foot in the door, you are being exposed to a professional engineering environment, and, most importantly, you are likely starting the process for your first security clearance, which gives you an incredible advantage.

Try targeting smaller subcontractors. Smaller, local defense subcontractors are often more flexible in their hiring than the giant Primes. They may not have a formal, structured internship program, but they are often in need of smart, capable young engineers to help with real, hands-on hardware tasks. This is a fantastic place to get your hands dirty, to work on real hardware, and to build a powerful story for your resume for when you apply to the Primes in your junior year.

The Multi-Pronged Attack

A successful internship hunt is a campaign fought on multiple fronts simultaneously. You cannot rely on a single source. The students who succeed are the ones who cast a wide, intelligent net.

Front #1: The University Career Fair (Your Highest Priority)

The university career fair is the Super Bowl of recruiting. For one or two days, the entire industry comes to you. It is a noisy, crowded, and often overwhelming environment, a gauntlet of long lines and brief conversations. But it is also your single best opportunity to make a human connection.

Reconnaissance: Weeks before the event, get the list of attending companies. Your goal is to create a prioritized target list of the 5-10 defense-related companies that will be present. For these top targets, go deep. Your reconnaissance must be specific:

> What do they do *at the location they are hiring for*? Know the difference between a company's radar division and its satellite division.
>
> What are their major public programs? Mentioning a specific program shows you've done your homework.
>
> What are their recent news and contract wins? A quick news search can give you relevant talking points.
>
> *The Elevator Pitch:* You must have a crisp, confident introduction ready.

The Formula:

1. *Introduction:* "Hi, my name is [Your Name]..." 2. *Value Proposition:* "I'm focusing on [Your Specialty], and I have hands-on experience from my [Project]..." 3. *The Connection:* "I was really interested in [Company Name]'s work on the [Specific Program]..."

Practice this out loud until it sounds natural.

Front #2: Direct Company Career Sites

Not all internships are posted on university job portals. Many excellent companies will not be at your career fair.

Expand Your Search: Use the "Domains" chapters to find other companies. Search for "naval engineering firms Virginia Beach" or "satellite manufacturers Denver Colorado." This is how you find the smaller subcontractors.

The Weekly Ritual: Create a spreadsheet of 15-20 companies. Every Friday morning during the fall semester, visit their career pages directly.

Front #3: The Alumni Network (Your Secret Weapon)

This is your most powerful tool for getting your foot in the door.

The Informational Interview: Your goal is not to ask for a job; it is to ask for 15 minutes of their time to learn from their experience.

What to Ask:

- "Could you tell me a little bit about your career path...?"
- "What is the culture like on your team...?"
- "What are some of the key skills or courses you took...?"
- "Based on my resume/experience, are there any areas you think I should focus on improving?"
- "Based on our conversation, do you have any advice on which groups or types of roles within [Company] you think I might be a good fit for?"
- "Is there anyone else you think it might be valuable for me to talk to?"

A Note on the Security Clearance:

A major reason for the early application timeline is the lengthy security clearance process. If you are selected, you will be sponsored for a clearance investigation. This will begin with you filling out an extensive online questionnaire called the SF-86. We will cover this in detail in a later chapter, but for now, know that complete and utter honesty is the only rule. This process can feel intimidating, but it is your single biggest advantage. Successfully completing an investigation as a student means you are already "in the system," making you an incredibly attractive and low-risk candidate for a full-time offer upon graduation.

The Performance

After a successful hunt, the real work begins. Landing the internship is only half the battle. Your performance during the summer is your three-month, high-stakes audition for a full-time role. The company is investing significant amounts of time, money, and resources in you, and they are watching closely. They are not just evaluating your technical skills; they are evaluating your potential. Your goal is not just to complete your assigned tasks, but to be so proactive, so engaged, and so effective that they cannot afford to let you go.

The Daily and Weekly Rhythm

One of the biggest adjustments from college to the professional world is learning to manage your own time. The internship is your first taste of a professional environment. The expectation is not to "punch a clock" for eight hours; it is to "get the mission done." This requires a new level of professional time management and proactive communication. Adopting a structured "battle rhythm" is the key.

Your Daily Rhythm

The first 30 minutes of your day should not be spent diving head-first into a technical problem. It should be spent "booting up." This means checking emails, reviewing your calendar for any meetings, and, most importantly, creating a simple, prioritized to-do list for the day in your notebook. You should block out 2-3 hour chunks of "deep work" on their calendar to focus on your primary project. This signals to others that you are busy and helps protect your concentration. You should spend the final 15 minutes of every day tidying up your work, saving your files, and, most importantly, writing a one or two-sentence summary of what you accomplished in a notebook. This is invaluable for the weekly report to your manager.

Your Weekly Rhythm

Monday morning is for establishing your plan for the week. This is when you should proactively align expectations with your mentor and manager. This will not only help you stay productive, but it will show them that you are organized and thinking ahead. Spend the last five minutes of your Friday, or the first five minutes of your Monday, reviewing your project plan. Prepare a simple list of your goals for the week and any questions or anticipated roadblocks you have.

On Fridays, you should be composing a status report. This is the single most powerful habit an intern can build as it documents your activities each week and can be important later on. Your report should be a concise, professional email to your manager and/or mentor every Friday afternoon.

"Weekly Status Email" Script:
Subject: Weekly Status Update - [Your Name] - Week of [Date]
Hi [Manager's Name],

Here is a quick summary of my progress this week on the [Project Name] project:

Accomplishments This Week: ...

Plan for Next Week: ...

No roadblocks at this time. Please let me know if you have any questions.

Thanks,

[Your Name]

This simple email makes you look incredibly organized and professional. It makes it easy for your manager to track your progress and to write your performance review.

Succeeding on the Job

Your summer internship can be thought of as a three-act play: the first month is about learning and earning trust, the second is about delivering results, and the third is about building your network and making a lasting impact.

The First Month - Earning Trust

Your first few tasks may seem simple, even trivial. They are not. They are a test of your diligence, your attention to detail, and your problem-solving process. During this time, your job is simply to be a sponge. Your goal in your first week is to be a professional, engaged, and humble learner. Show up with a physical notebook and write everything down. Inevitably, you will have periods of downtime. Do not ever sit at your desk staring at your phone. If you finish a task, ask your mentor if you can help with anything else.

The Art of the "Dumb" Question

One of the biggest fears for a new intern is the fear of looking stupid. They are often so terrified of asking a "dumb question" that they remain silent, which is far more dangerous. Senior engineers and managers do not expect interns to know everything. A silent intern is often seen as a disengaged or confused intern. Asking a foundational question is a sign that you are trying to learn. The problem is not the question, but how it's asked.

The "Context" Frame: Always provide the context of why you are asking and show that you've made an initial effort.

Bad Question: "What's a CDR?" (Lazy).

Good Question: "I heard you all talking about preparing for the CDR. I know it stands for Critical Design Review, but could you tell me what that means for our team specifically?"

The "Confirm My Understanding" Frame: This is a powerful technique for verifying information.

Instead of: "How does this work?"

Say: "Okay, let me see if I'm understanding this correctly. The data comes from the sensor, goes through the FPGA, and then is sent to the main computer. Is that the right data flow?"

The Second Month - Delivering Results

By the beginning of your second month, the initial fog of the new job has started to lift. You know your teammates' names, you understand the basic acronyms of your project, and you have successfully completed your first few tasks. You have begun to earn a measure of trust. Now, in the heart of the summer, your mission shifts from learning to executing. In this phase, you will execute the bulk of your technical work. It is your time to go deep, to focus, and to deliver a high-quality result on your primary project.

This is where you build your technical reputation. Your performance during this mid-summer grind is what will give your manager the evidence they need to say, "This intern is not just a student; they are a capable engineer who can deliver." You must take deep, personal ownership of your project. Treat it as if you are the sole person responsible for its success. Think ahead. Anticipate problems. Do not wait to be told what to do next. If you know a key piece of information is needed from another team, proactively reach out to them. If you see a potential risk to your schedule, develop a mitigation plan and present it to your mentor.

This is also the time to manage your energy and your time effectively. The initial excitement has worn off, and the end is still far away. This is where the "Daily and Weekly Battle Rhythm" we discussed becomes so critical. Use your "deep work" blocks to make steady, consistent progress on your project. Use your Friday status emails to keep your manager and mentor fully aware of your progress, your successes, and your challenges. This period is not about grand, heroic gestures; it is about the quiet, professional discipline of doing excellent engineering work, day in and day out. Your goal is to enter the final month of your internship with the bulk of your technical project complete, giving you the time and the credibility to focus on the final, crucial phase of making a lasting impact.

The Final Month - Building Your Network and Making an Impact

As you enter the final month of your internship, your focus must undergo another strategic shift. The bulk of your technical project work should be nearing completion. Now, in this final phase, you will transition from being a good "doer" to being a savvy "finisher" and a proactive "networker." This is where you go beyond just completing your task and work to solidify your reputation, expand your influence, and make a powerful case for a full-time return offer.

This is the time to build your network with a purpose. Your goal is to have numerous people outside of your immediate team who can recognize your work and be able to vouch for you and your impact. This is not about office politics; it is about professional relationship-building. Begin your informational interview campaign in earnest, as we will detail in the networking chapter. Set up 15-minute meetings with senior engineers or managers of teams you find interesting.

How to Ask: Send a simple, respectful email: "Hi [Name], My name is Alex, I'm the intern on the structures team. I'm really interested in the work your team does in thermal analysis. I was wondering if you might have 15 minutes to spare in the coming weeks to tell me a little bit about your work."

What to Ask: Ask them about their journey: "How did you get started in this group? What have been some of your favorite projects? What advice do you have for someone just starting out in this industry?" People love to share their stories, and it builds a genuine connection that makes you a known, respected name across the organization.

Simultaneously, your focus must be on delivering a flawless final product. This is not just about finishing your analysis or your code; it is about the professional handoff. Go above and beyond in your documentation. Comment your code meticulously. Organize your analysis files into a clean, logical folder structure. Write a "Read Me" file that explains your work to the next person who will pick it up. This level of professional polish is a massive differentiator.

Finally, you must prepare for your final presentation. As we will discuss, this is your closing argument. It is your last and best chance to showcase your work to a wider audience of managers and senior engineers. By rehearsing it, making it data-driven, and delivering it with confidence, you are leaving a final, lasting impression of competence and professionalism. By the end of this act, you will have transformed yourself from a temporary intern into an indispensable future colleague.

The Anatomy of an Internship Project: A Case Study

To make the experience of succeeding as an intern tangible, let's follow a fictional intern, "Jane," a junior electrical engineering student, through the entire arc of a typical 12-week summer project. This narrative will show how the principles of the "Three-Act Play" and the "Battle Rhythm" come together in a real-world scenario.

Weeks 1-2: The Tasking and the "Fog of the Unknown"

On her first day, Jane is assigned a mentor, "Dave," a senior RF engineer. Her project, assigned by her new manager, sounds deceptively simple: "Characterize the performance of this new power amplifier and write a test report." Jane feels a mix of excitement and terror. She recognizes the amplifier from her coursework, but she has no idea what program it is for or what its specific requirements are. Her first week is not spent in the lab, but at her desk, reading. Dave gives her a stack of documents: the vendor's detailed datasheet, the system-level requirements document for the radar system the amplifier is a part of, and the company's internal test procedure standards.

Jane spends hours in her new notebook, creating a glossary of program-specific acronyms and sketching out a block diagram of the test setup she will need to build. She feels lost in the sea of new terminology, but she is actively gathering the intelligence she needs to form a plan. At the end of the first week, she has her first one-on-one with her manager. Using her notes, she is able to ask a set of intelligent, clarifying questions: "I've reviewed the requirements document. It looks like the most critical performance metric is the amplifier's linearity at high temperatures. Is that the primary focus for my testing?" Her manager, impressed, confirms her understanding. Jane has survived the initial fog and now has a clear, prioritized objective.

Weeks 3-6: The Mid-Summer Grind

Jane is now in the lab, tasked with building her test setup with a network analyzer, a power supply, and various RF cables. This is where the real world collides with her academic knowledge. The first major problem hits: the spectrum analyzer is giving her bizarre, noisy readings that make no sense, and she can't replicate the performance numbers from the vendor's datasheet. This is a critical moment. Instead of immediately panicking and declaring that the part is broken, Jane applies the "Struggle First" rule.

She spends a full day methodically troubleshooting: she swaps out the RF cables, she recalibrates the network analyzer, she re-reads the equipment manual to ensure her settings are correct, and she writes a simple script to see if the noise is consistent over time. She develops a theory: a subtle impedance mismatch caused by a faulty adapter is creating a grounding loop. Only then does she go to Dave and say, "I'm seeing a lot of noise that doesn't match the datasheet. I've already checked the cables, the software, and recalibrated the VNA. My theory is that it's a grounding loop from this specific adapter. Am I on the right track?" Dave, impressed by her methodical, evidence-based approach, spends 10 minutes at the whiteboard with her, explaining a common grounding technique. They swap the adapter, and the problem is solved. Later that day, Jane makes a note in her engineering notebook about the grounding loop issue and the solution, ensuring she has captured the lesson for the future. She has not only fixed the issue, but she has also earned a new level of respect from her mentor for her professional troubleshooting process.

Weeks 7-9: The Push to Deliver

The testing is now going smoothly. Day after day, Jane runs the amplifier through its paces, collecting gigabytes of performance data under different temperature and power conditions. The project now shifts from hardware and the lab to software and analysis. Jane spends her time in MATLAB, writing scripts to parse the massive data files, plot the results in a clear and logical format, and compare them directly against the system's requirements.

During this process, she discovers something interesting and unexpected. The amplifier's performance is excellent at room temperature, but as it heats up, its linearity degrades significantly faster than the vendor's datasheet predicted. This is her "aha" moment. It is the key, critical finding of her entire summer, a piece of new knowledge that could prevent a major system failure in the field. The work now becomes about communicating this finding with undeniable proof. She creates clear, professional plots that overlay her test data with the requirement line and the vendor's promised performance, making the discrepancy obvious. She begins to structure her final test report, focusing not just on presenting the data, but on telling the *story* of the data.

Weeks 10-12: The Final Polish and the Professional Handoff

The core technical project is now complete, but Jane's work is not done. Her focus now is on her Final Presentation and creating a Handoff Package that will live on long after she leaves. She rehearses her presentation with Dave, who gives her critical feedback on how to frame her findings for a management audience. He tells her, "Don't just show them the problem; show them why it matters to the program and offer your recommendation."

But she also does something that truly sets her apart: she creates a perfect handoff package. She meticulously organizes all her raw test data into a clean, well-labeled folder structure. She goes through all of her MATLAB scripts, adding detailed comments to every section of the code so that another engineer can understand her process. And she writes a short "Read Me" document that explains exactly how to run her analysis scripts to reproduce her findings. She is ensuring that the engineer who picks up her work in September will be able to understand and build upon it instantly, without having to waste a week trying to decipher her work. This is the ultimate sign of a professional engineer, and it is the kind of act that gets remembered long after the internship is over.

The Final Presentation

Most internships culminate in a formal presentation where you brief your team, your manager, and often a wider group of department leaders on the work you've accomplished. This is your final exam, your closing argument. It is your last and best chance to solidify your reputation as a competent, professional, and high-potential engineer. This is not just a summary of your work; it is a demonstration of your communication skills and your professional polish. Do not treat it like a class project.

It's Not a Class Project: The audience is not a professor who is paid to listen to you for an hour. It is a group of busy managers and senior engineers who are taking time away from their own deadlines. Your presentation must be concise, data-driven, and ruthlessly focused on the results and the impact of your work. Aim for a crisp 15-20 minute presentation, leaving ample time for questions.

The "BLUF" Principle: You must adopt the military communication concept of "BLUF" (Bottom Line Up Front). Do not save your conclusion for the end. Your very first slide after the title page should be an executive summary that clearly states the project's goal, your key accomplishment, and your final recommendation. The rest of the presentation is simply the evidence to back up that claim.

Structure of a Great Presentation:

Title Slide: Your Name, Project Title, Your Mentor, Your Manager, and the Date.

Executive Summary (BLUF): In 3-4 bullet points, state the project's goal, your single most important finding/accomplishment, and your final recommendation. (e.g., "Goal: Analyze overheating in X unit. Result: Identified flawed assumption and designed a new air duct that reduced temps by 20°C. Recommendation: Proceed with manufacturing the new design for the prototype build.")

The Problem Statement: Briefly explain the problem you were tasked with solving and why it matters to the program.

Your Approach: Briefly explain the process you followed (e.g., "I developed a finite element model, ran a series of simulations under three different load cases, and validated the results against hand calculations.")

The Results: This is the heart of your presentation. Show the data. Use clear, well-labeled charts, compelling CAD models, or, if possible, a short video of your test. This is the evidence that proves your conclusion.

The Impact: How did your work benefit the program? Quantify it. ("This 20°C reduction in operating temperature provides a 50% margin against the component's operational limit, significantly reducing the risk of a mission-critical failure in the field.")

Lessons Learned & Future Work: What did you learn from the experience? What would be the next logical step in this project if it were to continue?

Thank You & Questions: End it cleanly and professionally, and be prepared to answer detailed questions about your work.

The Importance of Rehearsal: The day before your presentation, do a full, timed rehearsal with your mentor and perhaps another trusted teammate. Ask them for their honest, critical feedback. There is no excuse for going into your final presentation unprepared. A well-rehearsed, confident presentation is the final, lasting impression you will leave.

Year-Round Opportunities

A summer internship doesn't have to end in August. For students who have proven themselves to be valuable assets, many companies, especially those located near a university, offer year-round or "part-time" internships. This is an incredible and often overlooked opportunity that can transform your career trajectory before you even graduate.

This arrangement typically involves you continuing to work a reduced schedule, often between 10-20 hours per week, during the fall and spring semesters. You become a regular, integrated member of the engineering team, attending meetings, taking on real tasks, and continuing to learn and contribute. The flexibility is often significant; many managers will allow you to adjust your hours around your class schedule and exams.

The advantages of pursuing a year-round internship are immense. First, it allows you to build an even deeper, more lasting relationship with your team and your manager, solidifying your reputation as an indispensable contributor. Second, by being there for an entire year or more, you get to see a much larger portion of the engineering lifecycle, moving beyond a single summer task to contributing to a major design review or a critical test event. Finally, and most importantly, it makes the decision to give you a full-time offer upon graduation an absolute certainty. You are no longer an "intern"; you are a junior member of the team who just happens to be finishing their degree. If you perform well during the summer and your university is located near your internship site, do not be afraid to proactively ask your manager if this is a possibility.

The Aftermath - Securing the Return Offer

Your internship does not end on your last day when you turn in your badge. The final, professional steps you take in your last week and in the months that follow can be the difference between a great summer experience and a full-time job offer waiting for you upon graduation. The goal of this phase is to leave a lasting, positive impression, to formally express your interest, and to keep the lines of communication open, ensuring that all of your hard work during the summer translates into a secure future.

In your final week, you will likely have an exit interview with your manager. This is a formal opportunity for you to give and receive feedback. Be prepared. Your primary goal is to professionally and enthusiastically express your strong interest in returning to the company. You can say, "I have had a fantastic experience this summer, and the work this team does is exactly what I want to do in my career. I am very interested in returning for a full-time position after I graduate."

This is also your chance to show your maturity and commitment to self-improvement. Ask for feedback: "Is there any advice you have for me or any specific skills you think I should focus on in my final year of school to be an even stronger candidate when I come back?" This is an incredibly impressive question that shows you are thinking about your long-term growth.

Thank you notes are a small but incredibly important professional courtesy. On your last day, or the evening after, send a professional thank you email to your direct mentor, your manager, and a few of the key senior engineers or other colleagues you worked with. Thank them for their time, their guidance, and for the opportunity. It is a small act that leaves a huge, positive impression.

A massive and very common mistake interns make is disappearing for nine months after their internship ends. You have spent three months building a professional network; do not let it go cold. Set a reminder in your calendar. A few months into your fall semester, send your mentor and your manager a brief, professional email. This is not an "ask"; it is an update.

Example Script:

> "Hi [Manager's Name], I hope you're doing well. I was just thinking about my time at [Company Name] last summer and how much I learned from you and the team. I wanted to send a quick update on my progress this year. My senior design project is going well; I'm leading the software sub-team for our autonomous rover, and we just had a successful preliminary design review. I'm also taking a graduate-level course on AI that I'm really enjoying. Thank you again for the incredible internship experience. The lessons I learned about [mention a specific lesson] have been invaluable."

This simple act keeps you top-of-mind when the full-time hiring decisions are being made in the spring and shows that you are continuing to grow and develop your skills.

What If I Don't Get a Return Offer?

This is a common fear, and it's important to address it head-on. You can be a stellar technical performer, a proactive networker, and a model professional; you can do everything right and still not receive a return offer for a full-time position at the end of the summer. If this happens, it is crucial to understand that this is very rarely a reflection on you or your performance.

It's not a personal failure. Return offers are often dependent on program budgets and hiring forecasts that are set months in advance, long before you even start your internship. A major program might lose its funding, a contract might be delayed or canceled, or a department's hiring "headcount" for the next year might be frozen by senior leadership for reasons that have absolutely nothing to do with you. The business realities of the defense industry are complex and can change quickly. Do not interpret a lack of an offer as a personal failure.

You have still achieved a massive success. Even without a return offer, you have achieved a massive victory. You now have a relevant and powerful experience on your resume that makes you an incredibly attractive candidate to every other company in the industry. You have spent a summer learning the language, the culture, and the processes of the defense world. You are no longer a student with only academic projects; you are a proven commodity.

You can leverage your "golden reference." The most valuable asset you walk away with is not just the line on your resume, but your network. You now have a manager and a mentor who, if you did a good job, can act as a "golden reference" for your applications to other companies. A positive, detailed recommendation from a manager at a major Prime Contractor who can speak to your work ethic and your technical skills is an incredibly powerful asset that will immediately put your application at the top of the pile at a competing company.

Remember to play the long game. Your career is a marathon, not a sprint. Maintain the professional relationships you built during your internship. Send the thank you note. Connect on LinkedIn. The defense industry is a small and tight-knit world. The manager you worked for as an intern might be a director at a different company five years from now, and that positive connection could be the key to your next great opportunity. View your internship not as a simple pass/fail test for a return offer, but as the successful establishment of your first, critical foothold in the professional world.

From Intern to Asset

The internship gauntlet, from the first chaotic career fair to your final, polished presentation, is a transformative experience. It is the bridge between the theoretical world of your university and the applied, high-stakes world of professional defense engineering. It is a long, demanding, and often stressful process, but it is also the single most valuable investment you can make in your own future.

By following the playbook laid out in this chapter, you have done more than just learn how to get an internship. You have learned a set of professional skills that will serve you for your entire career. You have learned how to conduct professional intelligence, how to articulate your own value, how to build a network from scratch, how to take ownership of a project, and how to perform with diligence and integrity in a professional environment. These are the skills that separate the successful engineer from the merely brilliant one.

Ultimately, the goal of an internship is to transform yourself in the eyes of the company. You start as an unknown quantity, a student, a temporary helper. Through your hard work, your proactive networking, and your professional maturity, you end as a known and trusted entity, a proven problem-solver, a future colleague. You end as an asset.

Whether you walk away with a return offer in hand or with a powerful new experience on your resume and a network of golden references, a successful internship is a victory. You have successfully navigated the single most important filter in the hiring process. You have gained your foothold in the industry. You are no longer just a student with potential; you are an engineer with proof.

Part 5

The Application Arsenal

Chapter 17

Decoding the Job Posting

Before you can write a single word of your resume, you must first critically understand the document you are responding to: the job posting. This is the single most common and most fatal mistake that students make. They treat all job postings as generic requests for an "engineering intern." They create one standard resume, a one-size-fits-all document they are proud of, and they send it everywhere, hoping that its brilliance will be recognized.

In the modern world of online applications, this is a recipe for absolute failure.

A job posting is not a suggestion; it is a detailed set of instructions. It is a cheat sheet, written by the hiring manager, that tells you exactly what technical skills, software proficiencies, and professional traits they are desperately looking for. It is a direct communication from your future boss, but its true meaning is encrypted. Your job is essentially to analyze and decode this document to find the keywords that will unlock the path to an interview.

Why? Because your first reader is not a human. It is a machine. The Applicant Tracking System (ATS), the silent robot guardian of the corporate world, will be the first judge of your application. It is not impressed by your elegant prose; it is a simple keyword-matching algorithm. If your resume does not contain the specific keywords from the job description, the ATS will assign it a low relevancy score, and it will be silently archived in a digital purgatory. A human will never even see it.

This chapter is your guide to defeating the robot. It will teach you how to dissect a defense internship posting to find the critical information you need. We will teach you how to read between the lines, how to differentiate a non-negotiable demand from a simple wish, and how to build the "Keyword Blueprint" that will serve as the foundation for a resume that is impossible for both the machine and the human to ignore.

How to Read a Job Posting Like a Hiring Manager

The first skill you must develop is learning to differentiate between a hard requirement and a soft preference. Students often read a job posting like a legal contract, assuming that if they don't meet 100% of the criteria, they are automatically disqualified. This is a massive error in judgment that causes countless qualified candidates to needlessly take themselves out of the running.

A job posting is not a contract; it is a wish list. The hiring manager, often with input from Human Resources, is describing their ideal, perfect, "unicorn" candidate. They know this person rarely exists. They are looking for the candidate who is the *best fit*, not necessarily the *perfect fit*. Your job is to confidently assess if you are a best fit, and that starts with understanding the anatomy of the requirements.

The Hard Requirements

These are the true gatekeepers. They are almost always binary, you either meet them or you don't, and are often driven by legal or contractual obligations that the hiring manager has no power to waive. Wasting time applying for a role where you don't meet these is an exercise in futility.

U.S. Citizenship: For 99% of defense roles, this is an absolute, legally mandated requirement for security clearance eligibility and ITAR compliance. If it says "U.S. Citizenship required," it is not flexible. There is no gray area here.

Security Clearance Level (for experienced roles): If a non-entry-level job says "Active Secret Clearance Required," they mean it. The program has an immediate need for a cleared individual, and they cannot wait the 6-18 months it takes for a new clearance to be processed. As a student, you will rarely see this, but it is important to understand.

Specific Degree: If the role is for an RF engineer and it requires a Bachelor of Science in Electrical Engineering (BSEE), your mechanical engineering degree will not be a substitute. The fundamental coursework is too different.

The Soft Requirements

This is where the art of interpretation comes in. Most of the "requirements" in a job posting fall into this category. They are negotiable, and a strong profile can easily overcome a deficit in one area.

Years of Experience: This is the most misunderstood requirement. It is often a proxy for maturity and skill level, not a literal demand. If a role asks for "2 years of experience" but your project work and internships are highly relevant and you have demonstrated deep expertise, you should absolutely apply. A hiring manager will always choose a candidate with one year of perfect, relevant internship experience over a candidate with two years in an irrelevant field.

"Preferred" Skills/Software: This is the manager's dream list. If you have five "preferred" software skills listed and you are an expert in two of them and have a passing familiarity with a third, you are a strong candidate. The goal of this list is to attract the best possible talent pool. Do not self-disqualify because you haven't used every single tool.

Advanced Degrees: Especially for entry-level roles, a "Master's degree preferred" line is often included to attract high-achieving candidates, but it is rarely a hard requirement if a bachelor's degree is listed as the minimum qualification.

The Golden Rule: If you meet the hard requirements and can confidently say you have a strong foundation in about 60% of the listed "preferred" skills and responsibilities, you are a qualified candidate and you should apply. Let the hiring manager be the one to say no; never do their job for them by self-selecting out.

Decoding and Defeating the ATS

When you click "submit" on an online application, your resume does not go to a hiring manager's inbox. It is routed into a massive digital database managed by an Applicant Tracking System (ATS). This software is the robot guardian of the corporate world. Its primary job is to make the recruiter's life easier by filtering through the hundreds, sometimes thousands, of applications a single internship posting can receive. It does this in a very simple way: *it plays a game of keyword matching.*

The ATS scans your resume for specific keywords and phrases that the recruiter has pulled directly from the job description and tells it to look for. It then assigns your resume a relevancy score based on how many of those keywords it finds. If your score is too low because you described your experience using different words than the ones in the job posting, your resume is automatically archived in a digital purgatory. A human will never see it, no matter how brilliant you are or how perfect your experience is. This acts as the great resume killer—and a silent one. You will never know you failed the test; you will just experience the silence of the "resume black hole."

Your first mission, therefore, is not to impress the human; it is to satisfy the machine. This is not about cheating; it is about translation. It is about taking your authentic experiences and translating them into the specific language the ATS is programmed to understand. This requires a methodical, intelligence-driven process of decoding the job posting.

Step 1: The Tools of the Trade and the Mindset

Before you begin, you must adopt the right mindset and assemble the right tools. Your mindset is crucial: you are not a student reading a syllabus. You are an intelligence analyst deconstructing an enemy communication. Every word in the job posting was chosen for a reason, and your job is to extract every single piece of actionable intelligence from this document. Do not skim. Do not assume. Read every line with a critical eye, asking yourself, "What is the specific skill or quality they are *really* looking for with this phrase?"

Your tools for this mission are simple but essential. Do not attempt to do this on a screen, where it's easy to get distracted or read too quickly. Print out a physical copy of the job description. The human brain interacts differently with a physical document, and the act of marking it up will force you to read more carefully. Arm yourself with two different colored highlighters. One color will be for "Technical Skills," these are the nouns, the software, the tools, the processes, the "what" of the job. The other color will be for "Professional Skills," these are the verbs and adjectives, the "soft skills," the cultural fit, the *way* they want you to work.

Step 2: The Intelligence Extraction

This is the core of the decoding process. Begin by reading through the "Responsibilities" and "Preferred Qualifications" sections once without your highlighters, just to get a feel for the role. Then, on your second pass, begin your meticulous highlighting. You are searching for clues, and every highlighted word is a vital piece of evidence that will help you build your case. This is not a quick task; it is a deliberate and focused analysis.

Example keywords you might find in a posting for a Mechanical Design Intern:

Technical Skills (Yellow Highlighter): SOLIDWORKS, Creo, GD&T, ANSYS, Teamcenter, Mechanical Design, Structural Analysis, Tolerance Analysis, Design for Manufacturability (DFM)

Professional Skills (Blue Highlighter): team-based environment, communication skills, presentation skills, problem-solving, documentation

Example keywords you might find in a posting for a Systems Engineering Intern:

Technical Skills (Yellow Highlighter): MATLAB/Simulink, Python, DOORS, Cameo, SysML, Systems Engineering Lifecycle, V-model, Requirements Management, Requirements Traceability, Verification and Validation (V&V), Trade Studies

Professional Skills (Blue Highlighter): work independently, strong analytical skills, document findings, collaborate with multi-disciplinary teams, present to customers

Step 3: Create Your "Keyword Blueprint"

After you have thoroughly marked up the job description, the next step is to consolidate your intelligence. Open a new document or a notebook page. Your task is to now create a master list of all the words and phrases you've highlighted, organized by category. This is your "Keyword Blueprint." It is the DNA of the perfect candidate for this specific role, and your resume must be structured to reflect this DNA. This is not just a list; it is a strategic document that will guide every single word choice you make in the next chapter.

Technical Blueprint for the Systems Engineering Intern:
> *Software/Tools:* DOORS, MATLAB/Simulink
>
> *Core Concepts:* Systems Engineering Lifecycle, V-model, Requirements Management, Requirements Analysis, Trade Studies, Control Systems, Robotics, Verification and Validation (V&V)

Professional Blueprint for the Systems Engineering Intern:
> *Teamwork:* multi-disciplinary team, collaboration
>
> *Personal:* strong analytical skills, problem-solving, work independently
>
> *Communication:* document findings, present technical information

SIDEBAR: Decoding in Action

Here is a snippet from a real Mechanical Engineering Internship posting, followed by the "Keyword Blueprint" an analyst would extract.

> Job Posting Snippet:
> *"...The ideal candidate will assist in the mechanical design and structural analysis of aerospace components. Responsibilities include creating and modifying 3D CAD models in Creo, performing Finite Element Analysis (FEA) to validate designs against requirements, and documenting results for design reviews. The intern will work in a fast-paced, team-based environment and must have strong communication skills."*

Extracted Keyword Blueprint:
> *Software/Tools:* Creo, (Implied: ANSYS/NASTRAN for FEA)
>
> *Technical Skills:* Mechanical Design, Structural Analysis, 3D CAD Models, Finite Element Analysis (FEA)
>
> *Professional Skills:* Team-based environment, Communication skills, Documenting results

Step 4: Ensuring Your Resume Gets Seen

Creating your Keyword Blueprint is the critical first half of the battle. The second half is ensuring that your resume, which you will build in the next chapter, successfully incorporates this intelligence and is perfectly optimized to pass the filters. Simply writing the resume is not enough; you must put it through a rigorous, multi-stage validation process before you ever click "submit." This is your pre-flight check, your final quality control inspection, and it is non-negotiable. Skipping this step is like designing a critical part without ever running an analysis on it.

There are three key checks in this validation gauntlet: The AI Check, The Visual Check, and The Human Check. These checks are designed to test your resume against its three different audiences: the robot guardian (ATS), the busy recruiter (the six-second scan), and the discerning hiring manager (the deep dive). By passing all three, you can be confident that your application is in the best possible shape to get in front of a real human hiring manager and make a powerful first impression.

The AI Check: Fighting Fire with Fire

Since your resume's first reader is a machine, why not use a machine to check your work? You are about to face a robot gatekeeper. The amateur fights the machine; the professional *uses* the machine. In the past, this was difficult, but the rise of modern AI tools has given you a powerful new capability: the ability to run a simulation. You can now use AI to simulate the ATS screening process and get instant, valuable feedback. This is a crucial step that most of your competition will not take.

How to do it: Use a powerful AI chatbot (like Gemini Pro or Claude Sonnet). Your goal is to give it a specific role and a clear set of instructions. Copy and paste the following into your AI tool:

> *"Act as a professional resume reviewer and an expert-level Applicant Tracking System (ATS). I will provide you with two pieces of text: a Job Description and my Resume. Your mission is to perform a detailed analysis and provide feedback in four distinct parts:*
> *Part 1: Keyword Extraction.*
> *Analyze the Job Description and extract a list of the top 10-15 most important keywords. Categorize them into "Technical Skills" (software, tools, specific engineering concepts) and*

"Professional Skills" (soft skills, collaboration, communication).
Part 2: Resume Keyword Match Analysis.
Scan my Resume and identify which keywords from your Part 1 list are present and which are missing.
Part 3: ATS Score and Rationale.
Provide a percentage score from 0% to 100% representing how well my resume is optimized for this specific job description's keywords. Provide a brief 1-2 sentence rationale for this score based on the match analysis.
Part 4: Actionable Improvement Suggestions
Provide 3-5 specific, actionable suggestions for how I can improve my resume's alignment. For each suggestion, quote a phrase from my current resume and show me exactly how you would rewrite it to incorporate a missing keyword or to better reflect the language of the job description.
Here is the Job Description: [Paste the entire job description here]
Here is my Resume: [Paste the entire text of your resume here]"

This process gives you an instant, unbiased analysis. The AI's feedback in Part 4 is the most valuable part; it will give you specific, word-for-word examples of how to better align your experience with the language of the posting, giving you the secret weapon for optimizing your resume's content.

The Six-Second Skim Test

Once your content is optimized for the robot, you must ensure it is optimized for the human. A recruiter will give your resume a first scan that lasts, on average, only six to ten seconds. In that time, they are not reading; they are pattern-matching, looking for key information in specific places. Your resume must be incredibly clean, professional, and easy to skim.

How to do it: Print out your final, two-page resume. Pin it to a wall or lay it on a table. Step back about five feet. Now, glance at it for just six seconds. What jumps out at you?

Is your name and contact information immediately obvious?
Is your "U.S. Citizen" declaration prominent and easy to find?
Can you quickly locate the "Skills Summary" section and see the block of keywords?

Are your "Projects" and "Experience" sections clearly delineated with bold titles?

Does it look like a dense, unreadable wall of text, or is there a good balance of white space?

Are your bullet points concise and easy to scan?

If your resume looks cluttered or confusing from a distance, it will fail the human test. This visual check forces you to see your resume not as the author who knows where everything is, but as a busy recruiter who is seeing it for the first time.

The Human Check: A Multi-Layered Review

The final and most important check is a human one. You are too close to your own resume to see its flaws; you will read what you *meant* to write, not what you actually wrote. Getting fresh eyes on your document is non-negotiable, and the best strategy is a multi-layered one, moving from your peers to professional and industry experts.

Layer 1: The Peer Review

This is your first line of defense. Find one or two other ambitious engineering students, perhaps from your senior design team, and form a small "resume review" group. Make a pact to be brutally honest with each other and to review each other's tailored resumes before you apply for important positions.

The Goal: To catch the obvious errors. Check for typos and grammatical mistakes; these are instant killers of credibility. Check for clarity and impact. Ask your reviewer, "Based on this bullet point, can you tell me exactly what *I* did?" If they are confused, you need to rewrite it.

Layer 2: Leveraging University Resources

Your university provides a wealth of free resources designed for this exact purpose. Do not graduate without using them.

The Career Center: Your university's career center is staffed with professionals who have reviewed thousands of resumes. Many offer one-on-one resume review sessions or host resume workshops. While their advice may sometimes be more generic, they are experts at ensuring your format is professional, your language is strong, and that you are avoiding common mistakes. This is an excellent check for overall professionalism.

Your Professors: Your engineering professors, especially those with industry experience or those who advise the senior design projects, can be an invaluable resource. A professor who teaches a controls class, for example, can give you expert feedback on how you've described your controls-related project work, ensuring you are using the technical language correctly and impressively.

Layer 3: The Professional Review

This is the highest and most valuable level of review, and it is the one most students are too intimidated to seek out. Getting your resume in front of someone who actually works in the defense industry is a game-changer. They know exactly what their company looks for, and they can provide insights that no one else can.

How to Get It: This is where the work you did in the Alumni Network pays off. The informational interview is the perfect gateway to this. After a positive 15-minute conversation with an alumnus, you can make a polite request: "This has been incredibly helpful. As a final piece of advice, would you be willing to take a quick, two-minute glance at my resume and give me your honest first impression? I would be incredibly grateful for an insider's perspective." Most professionals, especially fellow alumni, will be happy to do this.

Why It's So Powerful: An industry professional can give you feedback that is impossible to get anywhere else. They might say, "My group is really focused on MATLAB scripting right now, you should make that more prominent," or "We don't use that software anymore, you should emphasize this other skill instead." And in the best-case scenario, if your resume is strong and they are impressed by your professionalism, they might say, "This looks great. Actually, my manager is looking for an intern right now. If you'd like, I can forward this on to them directly." This is how you bypass the online portal entirely and get your resume placed on the top of the pile.

From Decoding to Dominance

The process of decoding a job posting is the single most overlooked and most powerful strategy in the entire job hunt. It is the intelligence work that separates the prepared candidate from the hopeful amateur. By treating the job description not as a casual list but as a formal set of requirements, you shift your entire mindset from "applying" to "engineering a solution."

You have learned how to distinguish the true dealbreakers from the wish list, giving you the confidence to apply for roles that others might shy away from. You have learned how to meticulously extract the technical and professional keywords that serve as the password to get past the ATS. And you have learned a rigorous, multi-stage validation process that incorporates AI, visual checks, and a hierarchy of human reviewers to ensure your final product is flawless.

You are now armed with a Keyword Blueprint. This blueprint is the key to the entire application process. It is the raw intelligence you will use to forge your resume and cover letter into a targeted, compelling, and ultimately undeniable argument for why you are the right person for the job. In the next chapter, we will take this blueprint and use it, section by section, to construct your single most important application document: your keyword-optimized resume. You are no longer guessing; you are executing a plan.

Chapter 18

The Keyword-Optimized Resume

You now have your Keyword Blueprint. You have extracted the intelligence from the job posting and understand the specific skills and qualities the hiring manager is looking for. Now, it is time to forge your weapon. Your resume is the single most important document in your job hunt. It is your ambassador, your technical specification sheet, and your first impression all rolled into one. A great resume doesn't just list your accomplishments; it tells a compelling, evidence-based story that proves you are the solution to the hiring manager's problem.

This chapter will teach you how to build a powerful, "federal-style" master resume, optimized to please both the robot guardian (the ATS) and the human recruiter. We will go section by section, providing not just a template, but the strategic thinking behind each component. This is not just about formatting; it's about engineering a document that is designed to get you an interview.

The Unwritten Rule of the Two-Page Resume

One of the most persistent and damaging pieces of advice given to students is that a resume *must* be one page. That advice is for non-technical roles or for students with absolutely no project or internship experience. For an engineering student targeting the defense industry, a one-page resume is not a sign of efficiency; it is often a sign of inexperience. Your resume should be two pages long. Why? Because the hiring managers you are trying to impress are accustomed to the multi-page federal resume format. They are not looking for a brief summary; they are looking for evidence. They want to see the rich, detailed, quantifiable proof of your skills and the depth of your relevant coursework.

A two-page resume gives you the necessary space to provide the keyword-rich detail required to pass the ATS and, more importantly, to impress a discerning hiring manager. Do not feel the need to cram everything onto one page. Give your accomplishments the space they deserve. However, this is not a license to add fluff. Every single line must earn its place. A brilliant one-page resume is infinitely better than a bloated two-page resume filled with irrelevant information like "proficient in Microsoft Word." Having a one-page resume is not a problem, nor is it a limit.

Building Your Resume, Section-by-Section

We will now build your master resume. This will be your personal database, a comprehensive document containing *everything* you've ever done. For each job application, you will take a copy of this master document and perform a quick, 5-minute tailoring edit to align it with the specific keywords of the role.

1. Contact Information & Citizenship

The Intent:

The purpose of this section is simple: *to make it incredibly easy for a recruiter to know who you are, how to contact you, and whether you meet the most basic hiring requirement for the industry.* It must be clean, professional, and unambiguous. This is the letterhead of your personal brand; it sets the tone for the entire document.

How to Format It:

This section should be a clean, centered block at the very top of your first page. Use a slightly larger, bold font for your name to make it stand out. Use pipes (|) or other simple separators to make the contact line easy to read.

<u>Example Format:</u>

Jane Doe

Anytown, USA 12345 | (555) 123-4567 | jane.doe.eng@email.com |
linkedin.com/in/janedoe

U.S. Citizen

Common Pitfalls to Avoid:

An Unprofessional Email: An email address like
skaterboi98@email.com is an instant red flag. Create a simple,
professional address for your job hunt (firstname.last-
name@email.com).

A Casual Voicemail: If a recruiter calls and gets a casual or joke
voicemail greeting, it shows a lack of professional maturity. Record a simple, clear greeting: "You have reached the voicemail of
Jane Doe. Please leave a message."

Hiding Your Citizenship: Do not make the recruiter hunt for this information. By placing "U.S. Citizen" clearly and prominently at
the top, you are immediately signaling that you understand the
industry and meet its number one prerequisite.

2. Education

The Intent:

This section establishes your core credentials and academic foundation. For a student, it is one of the most important sections on the resume, sitting right at the top. It needs to be detailed and keyword-rich,
proving that you have the necessary theoretical knowledge for the role.

How to Format It:

Keep it clean and organized. The name of the university should be
bold, followed by the degree and graduation date on the next line. Use bullet point for your "Relevant Coursework" to make it a scannable, keyword-
dense list.

<u>Example Format:</u>

State University – University Park, USA

Bachelor of Science in Mechanical Engineering – Expected May 202X

Minor in Computer Science

Overall GPA: 3.75 / 4.0

- **Relevant Coursework:** Control Systems, Mechanical Vibrations, Heat Transfer, Introduction to Finite Element Analysis, Introduction to Robotics, Materials Science, C++ for Engineers

Common Pitfalls to Avoid:

Listing High School: Once you are in college, your high school is no longer relevant and takes up valuable space.

Vague Coursework: Simply listing "Engineering Electives" is a wasted opportunity. You must list the full, formal names of your most impressive technical courses. Use your Keyword Blueprint to choose which courses to highlight.

Lying About GPA: Never lie or round up your GPA in a misleading way. The company will verify it, and any discrepancy is an instant disqualification for falsification.

3. Skills Summary

The Intent:

View this section as your primary weapon for defeating the ATS and for impressing the human recruiter during their initial six-second scan. This is where you will deploy the intelligence you gathered in your 'Keyword Blueprint.' It is a dense, highly scannable block of keywords that acts as a quick summary of your technical capabilities. It should be comprehensive but perfectly organized.

How to Format It:

Use clear, bold subheadings to break the section into logical categories. Within each category, list your skills. This format allows a recruiter's eye to immediately jump to the category they care about most (e.g., "Software").

The 5-Minute Tailoring Rule:

Before you apply for a specific role, take five minutes to customize this section. Pull up the Keyword Blueprint you created for that job. Ensure that the top 3-5 technical keywords from the job description appear prominently in your Skills Summary. If the job description repeatedly mentions 'MATLAB' and 'Systems Engineering,' those terms must be in this section. If you listed 'Creo' on your master resume but this specific job calls for 'SOLIDWORKS,' swap them. This quick tailoring process dramatically increases your ATS score for each individual application.

<u>Example Format:</u>

SKILLS SUMMARY

Software & Tools: SOLIDWORKS (CSWP Certified), CATIA V5, ANSYS Mechanical, MATLAB, Simulink, Python (NumPy, Pandas), C++, LabVIEW, Microsoft Visio, Altium Designer

Technical Skills*:* Systems Engineering, Requirements Analysis, Finite Element Analysis (FEA), Geometric Dimensioning & Tolerancing (GD&T), Signal Processing, Circuit Design & Analysis

Lab & Shop Skills: Oscilloscope, Spectrum Analyzer, CNC Mill Operation, Manual Lathe, 3D Printing (FDM, SLA), Soldering & Rework

<u>*Common Pitfalls to Avoid:*</u>

Listing Obvious Skills: Do not waste space on skills that are assumed for any professional, like "Microsoft Word" or "Internet Browsing." This is a sign of a weak resume.

Being Vague: Don't just say "CAD Software." Say "SOLIDWORKS (CSWP Certified), Creo, CATIA V5." Be specific. The more specific you are, the more credible you sound and the more keywords you hit.

Disorganization: Do not list your skills in one giant, unreadable paragraph. The subcategories are essential for readability.

4. Projects

<u>*The Intent:*</u>

For a young, unproven engineer, the Projects section stands as the most important part of the resume while there is a lack of experience to populate your professional experience section. Your GPA shows you are a good student; your projects prove you are an engineer. This is where you provide the hard evidence of your capabilities. This section is not just a list; it is a collection of short, powerful stories designed to showcase your skills in action. Each bullet point should be a self-contained testament to your ability to solve a problem and achieve a result.

<u>*How to Format It:*</u>

Give each project its own clear entry. The project title should be bold and descriptive. The dates and your affiliation (e.g., "Senior Capstone Project") should be clear. The bullet points should be indented, creating a clean, hierarchical structure that is easy to read.

Example Format:

Autonomous Object-Tracking Turret | Anytown, USA

Senior Capstone Design Project – Aug 202X – May 202X

- Designed and modeled all structural components in SOLID-WORKS, performing FEA in ANSYS to validate the design against a 5g shock load requirement, achieving a 2.5 factor of safety.
- Developed 500+ lines of C++ code on a Raspberry Pi using the OpenCV library to process webcam imagery, achieving a 95% successful object tracking rate at 30 FPS.

Sidebar: The STAR Method and Quantification

The single biggest mistake students make in their resume is writing weak, passive bullet points that describe their duties instead of their accomplishments. To avoid this, you must adopt the STAR method. It is a simple but incredibly effective framework for structuring each bullet point as a compelling mini-story.

STAR stands for:

S - Situation: The context of the problem. (e.g., "On my senior design project...")

T - Task: The specific goal you were assigned. (e.g., "...I was tasked with designing the power system.")

A - Action: The specific, individual actions *you* took to achieve the goal.

R - Result: The quantifiable, positive outcome of your actions.

On a resume, you will combine these elements into a single, powerful bullet point, always starting with a strong action verb. The "Situation" and "Task" are often implied by the project title and the start of the bullet point. The real power comes from clearly stating your Action and the Result. Remember the mantra: Accomplishments, not responsibilities. Your resume should not be a list of what you were *supposed* to do; it should be a list of the quantifiable results you *actually* delivered.

Figure 5: The STAR Method. A framework for structuring compelling, evidence-based stories for resumes and behavioral interviews.

The second half of this powerful combination is *Quantification*. Numbers are the language of engineering, and they are the currency of credibility. An unquantified statement is an opinion; a quantified statement is a fact. Fill your resume with facts. They are unambiguous, credible, and they immediately show a recruiter the *scale* and *impact* of your work. You must make an almost obsessive effort to add a number, a metric, or a specific data point to every single bullet point if possible.

Example 1: The Mechanical Engineer

Weak: "Was responsible for designing the chassis for the rover."

Gold Standard: "Designed and modeled a lightweight aluminum chassis in SOLIDWORKS, and performed FEA in ANSYS to validate the design against a 3-foot drop requirement, achieving a 2.5

factor of safety while reducing total frame weight by 15% compared to the initial concept."

Example 2: The Electrical Engineer

Weak: "Worked on the power board for the project."

Gold Standard: "Designed and fabricated a 4-layer custom PCB in Altium Designer to regulate power for all motors and sensors, resulting in a 20% reduction in system wiring complexity and improving the vehicle's endurance by 10 minutes."

Example 3: The Software Engineer

Weak: "Wrote code to track objects."

Gold Standard: "Developed 500+ lines of C++ code on a Raspberry Pi using the OpenCV library to process a 1080p webcam feed, implementing a color-based filtering algorithm that achieved a 95% successful object tracking rate at 30 FPS."

5. Professional Experience (and Internships)

The Intent:

Functioning similarly to the Projects section, this area carries the added weight of a professional environment. It proves you can apply your skills in a real-world setting with real-world consequences. The exact same principles apply; use powerful, STAR-method bullet points to describe your accomplishments, not just your duties.

How to Format It:

The format should be identical to the Projects section to maintain a consistent and professional look. Clearly list the company, your title, and the dates of your employment.

Common Pitfalls to Avoid:

Listing Duties, Not Accomplishments: The hiring manager already knows what an intern is *supposed* to do. They want to know what *you actually did.* Don't say, "Was responsible for testing." Use a powerful STAR-method bullet to say, "Developed a MATLAB script to automate the analysis of 200+ hours of test data, identifying a critical performance issue missed by previous manual reviews."

Violating Security: Be very careful about the level of detail you provide. Do not ever list classified program names or specific, sensitive performance details. Describe your work using unclassified, functional language. When in doubt, ask your former manager for guidance.

Mishandling Your Security Clearance: Your security clearance is a valuable credential. If you were granted a clearance (even an Interim) or had an investigation initiated during an internship, you must list it. However, it must be done correctly.

Why It Matters: Using the correct terminology shows you understand the system. Claiming you "have" a clearance when you only had an investigation started is a common mistake and a red flag for dishonesty. Listing it correctly is a massive signal of your experience and value.

Example of listing your security clearance:

How to List It: Add a line item directly under your Contact Information & Citizenship section.

Correct Phrasing (if granted): "Security Clearance: Active Secret, Investigation Date: [Month, Year]"

Correct Phrasing (if in process): "Security Clearance: Secret Eligibility, Investigation Initiated: [Month, Year]"

6. Leadership & Activities

The Intent:

Use this section to showcase the "professional skills" you identified in your Keyword Blueprint. It proves you are a well-rounded candidate with experience in teamwork, leadership, and time management.

How to Format It:

Use a similar format to your Projects section. List the organization, your title, and the dates. Use a bullet point or two to describe your role or accomplishments, quantifying them if possible.

Example Format:

American Institute of Aeronautics and Astronautics (AIAA)

Team Lead, Design/Build/Fly Competition – Aug 202X – May 202X

- Managed a 10-person student team through the full design lifecycle of a competitive RC aircraft, resulting in the team placing 15th out of 110 international competitors.

Common Pitfalls to Avoid:

Listing Passive Memberships: Simply stating "Member, ASME" is not impressive. What is impressive is a leadership role.

Including Irrelevant Hobbies: Your car restoration project belongs in a "Personal Projects & Skills" section, not here.

A Note on the Samples

The following pages provide two sample resumes. The first is for an "Internship Hunter": a student in their sophomore or junior year whose experience is primarily project-based. The second is for a "Full-Time Candidate": a graduating senior with a strong internship on their record.

These are purely samples designed to illustrate the format and the principles we have discussed. Your resume must be longer if your experience warrants it. A graduating senior with two internships and a major capstone project will easily and justifiably fill two full pages with rich, detailed, and quantifiable accomplishments. Do not feel constrained by these specific examples; use them as a guide for structure, language, and the level of detail you should be striving for.

Jane Doe

Anytown, USA 12345 | (555) 123-4567 | jane.doe.eng@email.com |
linkedin.com/in/janedoe

U.S. Citizen

EDUCATION

State University – University Park, USA

Bachelor of Science in Mechanical Engineering – Expected May 202X

Minor in Computer Science

Overall GPA: 3.75 / 4.0

- **Relevant Coursework**: Control Systems, Mechanical Vibrations, Heat Transfer, Introduction to Finite Element Analysis, Introduction to Robotics, Materials Science, C++ for Engineers

SKILLS SUMMARY

Software & Tools: SOLIDWORKS (CSWP Certified), ANSYS Mechanical, MATLAB, Simulink, Python (NumPy, Matplotlib), C++, LabVIEW, Microsoft Office Suite

Technical Skills: Finite Element Analysis (FEA), Mechanical Design, Control Systems Theory, GD&T (ASME Y14.5), Data Analysis, Technical Writing

Lab & Shop Skills: 3D Printing (FDM), Manual Mill & Lathe, Soldering & Rework, Strain Gauge Application

PROJECTS

Autonomous Rover for Search & Rescue | Anytown, USA | Portfolio: [linktr.ee/janedoe/rover]

Senior Capstone Design Project – Aug 202X – Present

- Led a 4-person mechanical sub-team responsible for the chassis, drivetrain, and manipulator arm design for a university-sponsored robotics competition.
- Designed and modeled a lightweight aluminum chassis and a 2-DOF manipulator arm in SOLIDWORKS, performing FEA in ANSYS to ensure survivability against a 3-foot drop requirement.
- Developed a MATLAB script to process and visualize sensor data from the onboard LIDAR unit, improving the obstacle detection algorithm's reliability by 25%.

- Collaborated with the electrical engineering team to integrate motors, servos, and sensors, and co-authored the final 50-page design report and presentation.

Formula SAE Suspension Team | Anytown, USA

Student Design Competition – Aug 202X – May 202X

- Designed the bell crank and pushrod suspension geometry for the team's race car, optimizing for camber gain and roll resistance using vehicle dynamics simulation.
- Authored a detailed work instruction for the assembly and maintenance of the suspension system, reducing track-side adjustment time by 15%.

John Smith

Metroville, USA 54321 | (555) 987-6543 | john.smith.ee@email.com |

linkedin.com/in/johnsmitheng

U.S. Citizen

EDUCATION

Tech University – Big City, USA

Bachelor of Science in Electrical Engineering – Expected May 202X

Overall GPA: 3.68 / 4.0

- **Relevant Coursework:** Digital Signal Processing, RF & Microwave Circuits, Embedded Systems, Control Systems, Communication Systems, FPGA Design

SKILLS SUMMARY

Software & Tools: MATLAB, Simulink, C/C++, Python, VHDL, Altium Designer, SPICE, LabVIEW

Technical Skills: Digital Signal Processing (DSP), RF Circuit Design, Antenna Theory, Embedded C, FPGA Programming, Requirements Management (DOORS)

Lab & Shop Skills: Oscilloscope, Spectrum Analyzer, Network Analyzer, Soldering & Rework (SMD)

PROFESSIONAL EXPERIENCE

ABC Defense Corporation – Huntsville, USA

RF Engineering Intern – May 202X – Aug 202X

- Assisted in the design and testing of a new X-band receiver for a radar warning system, improving the noise figure of the test setup by 2 dB through careful component selection.
- Developed a MATLAB script to automate the analysis of over 200 hours of test data, identifying an intermittent performance issue that was not caught by previous manual reviews.
- Authored a 15-page test procedure for the new receiver, which was adopted as the new standard for the lab.
- Gained experience in requirements management by helping a senior engineer trace test results back to system requirements in the IBM DOORS database.

PROJECTS

Software Defined Radio (SDR) GPS Receiver | Big City, USA

Personal Project – Jan 202X – Present

- Designed and implemented a complete GPS receiver in a GNU Radio SDR framework using Python.
- Developed the DSP algorithms for signal acquisition, tracking, and decoding of the GPS navigation message, achieving a successful position lock from raw satellite signals.

Chapter 19

The Art of the Cover Letter

If the resume is the technical specification sheet that proves you are qualified, the cover letter is the narrative that proves you are interested. In a world of online applications and automated systems, it is a rare and valuable opportunity to speak directly to a human being, to bridge the gap between the cold, hard facts of your resume and the motivated, problem-solving engineer they are hoping to hire. A great cover letter doesn't just re-state your resume; it provides the *context* and the *passion* that your resume, by its very nature, cannot. If the resume is the engineering drawing, the technical specification sheet of your capabilities, then the cover letter is the executive summary. It is the one-page narrative that connects your skills to the company's mission and convinces the 'customer' why they should invest their time in reading the detailed specs.

Many students make one of two mistakes, both of which are fatal. The first is not submitting a cover letter at all, assuming it's an optional formality. This is a massive missed opportunity. In a stack of a hundred equally qualified resumes, the ten candidates who took the time to write a thoughtful, tailored letter are the ones who get the first look. It is the first and easiest way to signal that you are a serious candidate who is willing to do the work.

The second, and far more common, mistake is submitting a generic, boring template. It's an understandable temptation; you're busy, you're applying to dozens of places, and it feels like an inefficient use of time. A bad cover letter, one that is lazy, full of clichés, and clearly copied and pasted for a dozen different applications with only the company name changed, is worse than no cover letter at all. It is an act of professional disrespect. It tells the recruiter that you are not genuinely interested in their company, only in getting *a* job, any job. It signals a lack of attention to detail and a lack of effort, two qualities that are antithetical to the engineering mindset.

A great cover letter, however, can be the deciding factor that gets you an interview. It can overcome a slightly lower GPA or a lack of direct internship experience by telling a powerful story of your passion and your relevant project work. This chapter will teach you how to deconstruct and then construct a compelling cover letter, paragraph by paragraph. We will give you the tools to turn this often-dreaded task into one of your most powerful job-hunting weapons.

The Goal of the Cover Letter

Before you write a single word, you must understand the strategic purpose of the document. Your resume is brilliant at answering the "what;" what skills you have, what software you know, what projects you've done. It is a comprehensive but impersonal catalog of your capabilities. Your cover letter has a different, more strategic purpose. It must answer the three "whys" that are at the forefront of every recruiter's mind when they are sifting through a mountain of applications.

Why *this* company? Consider this your first and most important test. Out of all the great engineering companies in the world, why do you want to work here? The recruiter is looking for evidence that you have made a deliberate choice, not just a random click. A specific, well-reasoned answer proves you've done your research, you understand their place in the industry, and you are genuinely motivated by their mission. This is where you show that you are not just looking for a job, but for a place to build a career.

Why *this* role? What about this specific internship or job description caught your eye and made you feel that you were a perfect fit? This shows you've actually read and understood the posting. It proves that you are applying with intent, not just shotgunning your resume to every open position. It's your chance to say, "I see the problem you are trying to solve with this role, and I am the person who can help you solve it."

Why *you?* As the most critical question, this requires you to connect your experience to their need. Out of the hundred other qualified students who applied, why are you the best fit? This is your chance to highlight your single most relevant accomplishment and explicitly connect it to the responsibilities listed in the job description. Your resume lists all your accomplishments; your cover letter highlights the one that matters *most* for this specific job.

If your cover letter doesn't clearly and concisely answer these three questions, it has failed. It should not be a rambling summary of your resume; it should be the compelling, one-page executive summary that makes the reader eager to *then* go and read your resume in detail.

How a Recruiter Reads Your Cover Letter

To write a compelling letter, you must first understand the reality of the person reading it. A recruiter or hiring manager for a popular internship might have a stack of 100 applications. Their first goal is not to find the perfect candidate; it is to reduce the stack to a manageable size. They are in a ruthless filtering mode. Your cover letter is their first and fastest tool for this.

They read your letter with three simple questions in mind, placing your application into one of three piles:
The *"No" Pile:* This is the easiest pile to get into. It includes letters with typos or grammatical errors (shows a lack of attention to detail). It also includes the obviously generic, copy-and-pasted letters. These are seen as low-effort and disrespectful, and the application is often discarded without even looking at the resume. You have been filtered out.

The *"Maybe" Pile:* This is the default pile. The letter is competently written, it mentions the correct company and job, but it's generic. It says things like, "I am a hard worker and passionate about engineering." It's not bad, but it's not memorable. The recruiter will sigh, put it in the "maybe" pile, and then move on to the resume, hoping to find something more compelling there. You have survived, but you have not gained an advantage.

The *"Yes" Pile:* This is the magic pile, and it is the smallest. This is for the letters that answer the "Three Whys." The recruiter reads a specific sentence like, "...I have long admired your company's pioneering work on the Manta Ray UUV program..." and their brain immediately lights up. "This person did their homework. They are genuinely interested in us." They then read a sentence like, "...on my capstone project, I used MATLAB to develop a PID controller that reduced flight path deviation by 50%..." and they think, "This person has the exact skills we need for this role." This letter doesn't just get put in the pile; the recruiter will often flag it, star it, and read the attached resume with a sense of excitement and anticipation. Your letter has done its job. It has turned you from a number into a compelling candidate.

Your mission is not to write a letter that avoids the "No" pile. Your mission is to write a letter that puts you directly into the "Yes" pile. The rest of this chapter will show you how.

The Structure of Your Cover Letter

Forget the long, multi-page letters you may have seen in the past. A great cover letter for an engineering role is concise, professional, and gets straight to the point. It is an act of professional courtesy to the busy person reading it. It should be three, maybe four, paragraphs long. No more than a single page. The recruiter is a busy person; your letter must be powerful and immediately understandable.

Paragraph 1: The Hook - State Your Purpose and Specific Interest

As the most important paragraph, the hook has about ten seconds to convince the reader that the rest of the letter is worth their time. It must immediately tell them who you are, what you are applying for, and, most importantly, why you are specifically interested in them. This is where you deploy the intelligence you gathered from your job description decoding and your company research.

How to Write It: A Line-by-Line Breakdown

The Opening: Start with the specific role and the Job ID. This is a simple, professional courtesy that helps the recruiter immediately categorize your application.

Example: "I am writing to express my enthusiastic interest in the Systems Engineering Intern position (Job ID: *12345*) I found on your company's career portal."

The Credential: Immediately state who you are and what your specialty is. This frames your expertise.

Example: "As a third-year Mechanical Engineering student at [Your University] with a focus on control systems and robotics..."

The Connection (The Most Critical Sentence): This is your chance to show you've done your research. This must be specific and genuine.

Weak: "...I have always been impressed by ABC Defense's innovative work." (This is a generic and meaningless cliché that could apply to any company.)

Good: "...I have been following your company's work in unmanned systems, and I am very interested in your mission." (Better, but still a bit generic.)

Gold Standard: "...I have long admired [*Company Name*]'s pioneering work in autonomous undersea systems, particularly the Manta Ray program. The opportunity to contribute to a mission of such technical challenge and national importance is my primary motivation for applying." (Specific, names a program, and connects it to your personal motivation.)

Paragraph 2: The Pitch - Connect Your Best Story to Their Needs

Serving as the body of your letter, this section is where you make your case. Do not simply regurgitate your resume. A common mistake is to write, "As you can see on my resume, I have experience in SOLIDWORKS, ANSYS, and MATLAB." The recruiter can already see that on your resume, and it's their job to read it. Your goal here is to select your single most relevant project or accomplishment and tell a brief, powerful story about it.

This is not a guess. The "most relevant" project is the one that most directly aligns with the "Keyword Blueprint" you extracted from the job description in the previous chapter. Look at their stated needs, then pick the one story from your experience that proves you can meet those needs. Your mission is to explicitly connect your best accomplishment to the key requirements from the job posting, demonstrating that you are the solution to their problem.

How to Write It

Create a direct link to the job description: "The job description emphasizes a need for candidates with hands-on experience in modeling and simulation using MATLAB/Simulink." This shows you've read the posting carefully and are directly addressing their stated need.

Tell the story of your project: This is a prose version of a STAR-method bullet point.

> "In my senior capstone project, I was the lead for the Guidance and Control sub-team, tasked with developing the autonomous navigation system for our rover. I successfully used Simulink to model the vehicle's dynamics and then developed a PID controller in MATLAB that resulted in a 50% reduction in flight path deviation in our validation testing. This experience has given me a strong practical foundation in the exact tools and techniques your team is looking for."

This paragraph takes one bullet point from your resume and breathes life into it. It gives the recruiter a compelling narrative of you solving a relevant problem, showing not just that you *have* a skill, but that you have *successfully applied* it to achieve a quantifiable result. This is far more powerful than a simple list of skills.

Paragraph 3: The Close - Reiterate Interest and Call to Action

End it cleanly and professionally. You have made your case. Now, you need to summarize your interest and make it clear that you are eager for the next step.

How to Write It

Restate your enthusiasm and summarize your fit:

> "My hands-on project experience in autonomous systems, combined with my coursework in control theory, has solidified my desire to pursue a career in the defense industry, and I am confident I have the skills and drive to be a valuable contributor to your team this summer." This is your final, confident summary of why you are a great match.

The Call to Action:

> "Thank you for your time and consideration. My resume is attached for your review, and I look forward to the

opportunity to discuss my qualifications further in an interview."

This is a professional and standard closing that clearly states your desire for the next step.

<u>*Common Pitfalls and Final Touches*</u>

The Generic Template: The single biggest mistake is writing one cover letter and sending it everywhere. Recruiters can spot a generic template from a mile away. You *must* tailor the company name, the job title, and your "Why Company" statement for every single application.

Typos and Grammatical Errors: A single typo can get your application thrown out. It shows a lack of attention to detail, which is a cardinal sin for an engineer. Proofread it three times. Then have someone else proofread it.

Addressing it: If you can find the name of the university recruiter on LinkedIn, use it. "Dear Ms. Smith" is always better than the generic "Dear Hiring Manager."

File Format: Always save and send your cover letter as a PDF to preserve formatting. Name the file professionally (e.g., *FirstName_LastName_CoverLetter_ABC_Corp.pdf*).

Think of your application as the opening salvo in your campaign to win the job. The generic applicant fires a scattered, un-aimed volley and hopes for a lucky hit. The strategic applicant, however, understands the power of precision.

Your keyword-optimized resume is the hypersonic missile; a dense, data-rich, and incredibly powerful instrument designed to penetrate the automated defenses of the ATS. It proves, with overwhelming evidence, that you are qualified.

Your cover letter is the laser designator. It is the human element, the flash of light on the target that says, "Right here. This is the one that matters." It provides the context, the narrative, and the intent that guides the hiring manager's attention. It proves that you are not just qualified, but that you are interested.

By following the framework in this chapter, you have engineered a document that is targeted, respectful of the reader's time, and tells a compelling story. You have created the perfect complement to your resume; a one-two punch that will dramatically increase your chances of hitting the target: the interview.

A Note on the Samples

The following pages provide two sample cover letters. The first is for an "Internship Hunter": a student in their junior year who is leveraging their project work to land a key internship. The second is for a "Full-Time Candidate": a graduating senior who is using their internship experience to make a powerful case for a full-time role.

These are purely samples designed to illustrate the structure, tone, and level of specificity you should be aiming for. You should use them as a guide to build your own unique and authentic letter, tailored to your own experiences and the specific job you are applying for.

Sample Cover Letter 1: The Internship Hunter
Jane Doe

123 University Drive, Anytown, USA 12345

(555) 123-4567 | jane.doe.eng@email.com | linkedin.com/in/janedoe

[Date]

[Hiring Manager Name or "University Recruiting Team"]

[Company Name]

[Company Address]

Subject: Application for Mechanical Engineering Intern Position (Job ID: 98765)

Dear [Mr./Ms. Last Name or Recruiting Team],

I am writing to express my enthusiastic interest in the Mechanical Engineering Intern position (Job ID: 98765) I found on the [Company Name] career portal. As a third-year Mechanical Engineering student at State University with a focus on robotics and control systems, I have been following your company's pioneering work in unmanned systems with great admiration. The opportunity to contribute to a program as technically challenging and important as the Manta Ray UUV is my primary motivation for applying.

The job description emphasizes a need for candidates with hands-on experience in the design and analysis of complex electro-mechanical systems. In my senior capstone project, I am the lead for the 4-person mechanical sub-team responsible for designing the chassis and manipulator arm for an autonomous search-and-rescue rover. I have spent the last semester designing the rover's lightweight aluminum chassis and a 2-DOF manipulator arm in SOLIDWORKS, and subsequently used ANSYS to perform the structural analysis to validate the design against the competition's 3-foot drop requirement. This project has given me a strong, practical foundation in the exact design, analysis, and integration skills that your team is looking for.

My hands-on project experience, combined with my coursework in control systems and robotics, has solidified my desire to pursue a career in the defense industry. I am confident that my skills and my passion for solving complex, mission-critical problems would allow me to be a valuable contributor to your team this summer. Thank you for your time and consideration. My resume is attached for your review, and I look forward to the opportunity to discuss my qualifications further in an interview.

Sincerely,

Jane Doe

Sample Cover Letter 2: The Full-Time Candidate

John Smith

456 Tech Avenue, Metroville, USA 54321

(555) 987-6543 | john.smith.ee@email.com | linkedin.com/in/john-smitheng

[Date]

[Hiring Manager Name or "University Recruiting Team"]

[Company Name]

[Company Address]

Subject: Application for Entry-Level RF Engineer Position (Job ID: ABC-123)

Dear [Mr./Ms. Last Name or Recruiting Team],

I am writing to express my strong interest in the Entry-Level RF Engineer position (Job ID: ABC-123) posted through the Tech University career portal. As a graduating senior with a Bachelor of Science in Electrical Engineering and a successful RF engineering internship at ABC Defense Corporation, I have developed a deep passion for the challenges of designing and testing advanced radar and communication systems. I am particularly drawn to [Company Name]'s reputation as a world leader in electronic warfare, and I am eager to apply the skills I have already developed to a full-time role on a program of such national significance.

During my internship last summer, my primary project involved assisting in the design and testing of a new X-band receiver for a radar warning system. The job description for this role emphasizes a need for engineers who can analyze test data to identify performance issues. In my internship, I was tasked with analyzing over 200 hours of performance data from our prototype receiver. By developing a MATLAB script to automate the analysis, I was able to identify a critical, intermittent performance issue related to thermal noise that had been missed by previous manual reviews. This work not only gave me hands-on experience with network analyzers and spectrum analyzers, but also taught me the importance of a methodical, data-driven approach to problem-solving.

My internship experience, where I was able to directly contribute to a critical defense program, confirmed my desire to build a career in this industry. I am confident that my foundational knowledge of RF circuits, my proven ability to analyze test data, and my security clearance eligibility make me a strong candidate for this position. Thank you for your time and consideration. My resume is attached for your review, and I am eager to discuss how I can contribute to your team.

Sincerely,

John Smith

Part 6

Navigating the Great Filters

Chapter 20

Networking in a Closed World

Networking. For many engineering students, the very word is cringe-inducing. It conjures up images of schmoozing in ill-fitting suits, of forced, awkward conversations, and of desperately trying to "sell" yourself. It can feel like a game for business majors, a world away from the clean logic of a differential equation or the satisfying precision of a CAD model. But in an industry as formal and seemingly closed-off as the defense sector, understanding how to professionally connect with people is not just a soft skill; it is a critical requirement. The reality is that a personal connection, however brief, can be the single most powerful factor in getting your resume pulled from the digital pile and placed into the "interview" stack. In a world where hundreds of qualified students apply for every open position, a trusted referral from a current employee isn't just a foot in the door; it is the golden ticket that gets your resume hand-carried past the digital gatekeepers and placed on the hiring manager's desk.

This chapter is your demystification guide to this essential skill. We will reframe networking entirely. It is not about schmoozing or asking for a job. It is about gathering intelligence, making a professional impression, and building a bridge between you and the company. It's your chance to turn your name from a line of text on a screen into a face and a memorable conversation. We will provide a practical, step-by-step guide to the three most important networking arenas for a student: the high-intensity career fair, the professional world of LinkedIn, and the ultimate secret weapon: the informational interview.

The Psychology of a First Impression

Before we dive into the specific tactics of networking, we must first understand the fundamental psychology of what you are trying to achieve.

In any professional interaction, from a career fair conversation to an informational interview, the person across from you is on an intelligence-gathering mission, subconsciously asking themselves two primitive, critical questions:

"Can I respect this person?" *(Competence)*

"Can I trust this person?" *(Warmth)*

Every piece of advice in this chapter is designed to help you project both of these qualities simultaneously. Competence is your ticket to the game. It is the proof that you are a serious, intelligent, and capable engineer. Warmth is what makes someone want to help you. It is the proof that you are a respectful, curious, and trustworthy individual who would be a good colleague. A candidate who projects only competence can come across as arrogant. A candidate who projects only warmth can come across as friendly but unqualified. The magic happens when you project both.

Projecting Competence

Your competence is not just demonstrated by your GPA or your resume. It is demonstrated by your actions and your preparation.

Preparation is a form of respect. When you show up to a conversation having done your research, when you know the company's programs and you've read the person's LinkedIn profile, you are sending a powerful signal: "I respect your time enough to have prepared for this conversation." This is a massive sign of professional maturity.

Curiosity is a sign of intelligence. The best way to demonstrate your technical competence is not to brag about your projects, but to ask intelligent, insightful questions. An engineer is a professional question-asker. Asking a thoughtful question about a technical challenge the company is facing or a new technology they are developing proves that you are thinking at a deeper level.

Conciseness is a sign of clarity. Engineers are trained to be precise. Rambling, disorganized answers are a red flag. The ability to articulate a complex idea clearly and concisely, like in your elevator pitch, is a powerful demonstration of a clear and organized mind.

Projecting Warmth

Warmth is the often-overlooked half of the equation, and it is just as important. No one wants to work with a brilliant jerk.

Place the focus on them, not you. The single biggest mistake students make is to make the entire conversation about themselves. Your goal is to learn about *them*. People love to talk about their own experiences and share their hard-won wisdom. By making the conversation about their career, their projects, and their advice, you are not only gathering valuable intelligence, but you are also showing respect and building a genuine human connection.

Active listening is a superpower. Do not just wait for your turn to talk. Listen intently to their answers. Ask follow-up questions that show you were paying attention. ("That's fascinating that you worked on the early phase of that program. What was the biggest lesson you learned from that experience?") This makes the other person feel heard and respected.

Remember that gratitude is essential. Every interaction should begin and end with gratitude. "Thank you so much for taking the time to speak with me" is the most powerful phrase in your networking toolkit. It shows humility and respect.

As you read the rest of this chapter, view every tactic, every script, and every piece of advice through this lens. Ask yourself: "How does this action demonstrate both my competence and my warmth?" By mastering this dual projection, you will transform yourself from a student asking for a job into a future colleague that people are eager to help.

The University Career Fair

The university career fair is the single best networking opportunity you will have. For one or two days, the entire industry comes to you. It is a noisy, crowded, and often overwhelming environment, a gauntlet of long lines, brief conversations, and a thousand other students competing for the same few minutes of a recruiter's attention. But it is also your single best opportunity to make a human connection. Your goal is not to leave with a job offer; your goal is to leave such a lasting, positive impression that when your resume comes across the recruiter's desk a week later, they remember your face and your conversation.

The Weeks Before

Your work for the career fair begins weeks before the event. Get the list of attending companies from your career services office and create a prioritized target list of the 5-10 defense-related companies that will be present. For these top targets, go deep. Your reconnaissance must be specific.

What do they do *at the location they are hiring for*? A company like Northrop Grumman has dozens of sites. The one in Baltimore specializes in radar systems, while the one in Redondo Beach specializes in satellites. Know the difference.

What are their major public programs? Mentioning the "B-21 Raider program" to a Northrop recruiter or the "Aegis Combat System" to a Lockheed Martin RMS recruiter shows you have done your homework.

What are their recent news and contract wins? A quick search on a site like "Breaking Defense" or "Defense News" can give you incredibly relevant talking points.

The Elevator Pitch:

You must have a crisp, confident introduction ready. For this industry, your pitch must be a powerful fusion of your technical skills and your specific interest.

The Formula: 1. Introduction: "Hi, my name is [Your Name], I'm a junior Mechanical Engineering student here at [University]." *2. Value Proposition:* "I'm focusing on [Your Specialty], and I have hands-on experience from my [Project] where I [Specific accomplishment]." *3. The Connection (The Most Important Part):* "I was really interested in [Company Name]'s work on the [Specific Program you researched], and I'm hoping to find an internship where I can apply my skills."

Practice this out loud. Record yourself on your phone. It will feel awkward, but it is the only way to ensure it sounds natural and confident, not robotic.

Execution

The Approach: Do not go to your number one dream company first. The first conversation of the day is always the rustiest. Execute a 'Warm-Up': Approach a company from your second tier first to shake off the nerves and practice your pitch in a lower-stakes environment. As you approach a busy booth, 'Triage' the representatives: Is the person at the front a recruiter (HR) or a working engineer? The engineers are your primary target. They can have a real technical conversation and are often the ones who can directly refer you to their own hiring managers. Be polite to the HR reps, but try to maneuver toward a conversation with an engineer.

The Conversation: When it's your turn, make eye contact, offer a firm handshake, and deliver your elevator pitch. After your pitch, the most important thing you can do is ask an intelligent question.

Bad Question: "So, what kind of internships do you have?" (Generic and lazy).

Good Question: "I saw that your facility in Baltimore specializes in radar systems. I'm really passionate about RF engineering. What kind of projects do interns on those teams typically get to work on?"

The Close: After a few minutes, be respectful of their time. Say, "I know you have a lot of students to talk to, but I really appreciate your time." Ask, "What is the best way to formally apply for this role? Is there a specific requisition number I should use?" And crucially, ask for their business card. If they don't have one, ask for their name and email address.

The Digital Embassy

After the career fair ends, the networking campaign continues online. Your digital presence is a critical front in this battle. At the center of this world is LinkedIn. Most students treat LinkedIn as a simple, online resume; a static document that they only update when they are actively looking for a job. This is a massive missed opportunity. You must think of LinkedIn not just as a resume, but as your professional embassy. It is the place where you represent yourself to the professional world 24/7. More than that, it is an incredibly powerful listening post for gathering intelligence and a lighthouse for broadcasting your own value.

Building Your Embassy

Before you can use LinkedIn for outreach, your own profile must be a fortress of professionalism.

A Professional Headshot: This is non-negotiable. Get a clean, simple photo of you smiling, against a neutral background.

The Headline is Key: Your headline should not just be "Student at [Your University]." It should be your forward-looking career goal, packed with keywords.

> *Good Headline:* "Mechanical Engineering Student | Aspiring Aerospace & Defense Professional | SOLIDWORKS & ANSYS"

The "About" Section is Your Story: This is your chance to write a short, compelling narrative that connects your skills to your passions.

Detail Your Experience with STAR: Fill out your "Experience" section for your projects and past jobs using the same powerful, STAR-method bullet points from your resume.

LinkedIn as a Listening Post

This is the proactive, "always-on" part of your networking campaign.

Follow Your Target Companies: This will fill your feed with their press releases, product announcements, and cultural updates. This is how you learn that "Company X just won a new contract for a satellite program," a perfect talking point for a future interview.

Follow the Influencers and the News: Follow key industry news outlets and senior leaders at your target companies. A Chief Engineer might post about a technical challenge their team is facing. This is pure gold.

Use the Alumni Search Tool: LinkedIn's most powerful feature for a student is its university alumni tool. This is your primary tool for building a target list for the informational interviews.

LinkedIn as a Lighthouse: Broadcasting Your Value

This is the advanced move that will truly set you apart.

Engage Professionally: Find a post from a senior leader you admire. Leave a thoughtful, professional comment.

Weak: "Cool!"

Strong: "This is a fascinating application of MBSE. I'm currently studying systems engineering, and it's exciting to see how these digital tools are being used to accelerate development."

Share Your Work (Carefully): When you complete a major, non-proprietary project, write a short, professional post about it. Include a great picture or a short video. This turns your profile from a static resume into a dynamic portfolio.

The Alumni Network & The Art of the 'Ask'

The final front in your attack is the most powerful and the most underutilized: your university's own alumni network. While a cold message to a recruiter on LinkedIn might get ignored, a message to a fellow alumnus has an incredibly high chance of getting a positive response. You share a powerful, immediate bond, and most alumni are genuinely eager to help students from their alma mater. Your primary tool for leveraging this network is the informational interview. However, for many students, the hardest part is simply knowing what to say. This section is your playbook.

Scenario 1: The "Cold" Outreach to an Alumnus for an Informational Interview

This is the most common and most important "ask." Your goal is to be respectful, concise, and to make it incredibly easy for them to say yes.

The Script:

Subject: Question from a fellow [University Mascot] Engineer

Dear Mr./Ms. [Last Name],

My name is [Your Name], and I found your profile on the [University Alumni Network / LinkedIn]. I am a junior Mechanical Engineering student here at [University], and I am very interested in a career in the aerospace and defense sector.

As a fellow alumnus, I was hoping you might have just 15 minutes to spare in the coming weeks to share a bit about your experience at [Their Company]. I am particularly interested in your work in [Their Specific Field], and any advice you might have for a student hoping to follow a similar path would be greatly appreciated.

Thank you for your time and consideration.

Sincerely,

[Your Name]

[Link to your LinkedIn Profile]

Scenario 2: The Post-Career Fair Follow-Up

You had a good, brief conversation with a recruiter or an engineer. Your goal is to reinforce the connection. This should be sent within 24 hours.

The Script:

Subject: Following Up from the [University Name] Career Fair

Dear Mr./Ms. [Last Name],

It was a pleasure speaking with you yesterday at the [University Name] career fair. I particularly enjoyed our discussion about [Reference a specific topic you discussed].

As we discussed, I am very interested in the [Specific Internship Title] role. My experience on my senior design project, where I was responsible for [Your relevant task], aligns well with the challenges you described.

Thank you again for your time. As you suggested, I have formally applied online to Job ID: 12345. I look forward to the possibility of speaking with you again soon.

Best regards,

[Your Name]

Scenario 3: The Request for a Resume Review

You've had a successful informational interview. Now, you want to make a second, slightly bigger "ask." This should be done in a separate, follow-up email a day or two after.

The Script:

Subject: Quick Follow-Up Question

Dear Mr./Ms. [Last Name],

Thank you again for taking the time to speak with me on Tuesday. Your advice on [Reference a specific piece of their advice] was incredibly helpful.

I have one quick follow-up question, if you don't mind. As I begin to tailor my resume for applications, would you be willing to take a quick, two-minute glance at it and give me your honest first impression from an insider's perspective?

I have attached it below for your convenience. Thank you again for your mentorship.

Sincerely,

[Your Name]

Networking *During* Your Internship

While the career fair and online platforms are essential for getting your foot in the door, the single most powerful networking opportunity you will ever have as a student is during your internship. For three months, you are not an outsider trying to get in; you are an insider with a badge. You have been granted temporary access to the organization's most valuable asset: its people. Wasting this opportunity is a catastrophic mistake. Your goal is not just to do a good technical project, but to leave that summer with a network of 10-15 professionals who know your name, respect your work, and can act as your advocates for a full-time offer.

Your mindset during the internship must be one of proactive, professional curiosity. You are not just there to do your assigned task. You are there to learn about the entire organization, to understand how different teams work together, and to build relationships with the engineers and leaders who are doing the work that you find most interesting. This is not about "schmoozing" at the coffee machine; it is a deliberate campaign of intelligence gathering and professional connection.

Your Week-by-Week Guide

The First Month (Weeks 1-4): Build your base and earn the Right. Your initial focus should be on your immediate team. Flawlessly execute your initial tasks. Build a strong, positive relationship with your mentor, your manager, and your teammates. Your goal by the end of the first month is simple: your immediate team should see you as a reliable, hardworking, and pleasant addition.

The Second Month (Weeks 5-8): The informational interview campaign is the heart of your networking mission, which we will cover in more detail below.. Your goal is to conduct at least one, ideally two, informational interviews per week.

Start by asking your mentor or manager: "I'm really interested in learning more about the [thermal analysis] group. Is there a senior engineer on that team you think it would be good for me to talk to for 15 minutes?"

Once they give you a name, ask if they can make a "warm introduction" via email. If not, you have their permission to send a simple, respectful message yourself, mentioning their name in your opening line.

The Third Month (Weeks 9-12): Go for breadth, depth, and influence. You can now expand your reach.

Breadth: Go beyond your own engineering discipline. Ask the people you've met for recommendations: "Who is the sharpest software engineer you know?" This demonstrates systems-level thinking.

Depth: If you've discovered a passion for a specific area, ask for a follow-up meeting with the senior engineer you spoke to. Ask more detailed questions.

Influence: Politely ask your manager if they would be willing to introduce you to *their* manager for a brief 15-minute chat. This gets you on the radar of the people who make the hiring decisions.

The Informational Interview

The informational interview is your best way to build a bridge into the company. It can be a brief 15-minute conversation with a coworker at an internship that you initiate. The goal of this interview is to make it a conversation where the other person feel valued, not interrogated. Your questions should be open-ended and focused on their story and their expertise. Prepare 3-4 questions like those from this list:

"Could you tell me a little bit about your career path and how you got to your current role?" (People love to tell their story).

"What is the most interesting or challenging technical problem you're working on right now?" (Shows your interest in the actual work).

"What do you think is the most important skill for a new engineer to develop in their first few years at [Company Name]?" (Asks for valuable advice).

"Based on our conversation about my interest in [X], is there anyone else in the organization you think it would be valuable for me to speak with?" (This is the "ask" for a warm referral).

While the conversation should feel natural and warm, you, the strategic engineer, have three clear objectives to accomplish in your 15-minute call. These are your metrics for a successful mission.

Objective #1: Make a Genuine Connection. This is your primary goal. It is achieved by demonstrating your Warmth and Competence. Show up prepared (Competence). Be curious about their story (Warmth). Listen actively (Warmth). Ask intelligent questions (Competence). End with gratitude (Warmth). If they walk away thinking, "That was a sharp and impressive young person," you have succeeded.

Objective #2: Gather Actionable Intelligence. You are on a reconnaissance mission. You are trying to gather information that you cannot find on the company website. Your goal is to leave the call with at least one key piece of "ground truth" that will help you in your job search. This could be a piece of advice ("My group is really focused on MATLAB scripting right now, you should make that more prominent on your resume"), a key insight into the culture ("We're a very hands-on team; hiring managers really value project experience over GPA"), or an understanding of a real problem they are facing.

Objective #3: Get a "Warm Lead" to Your Next Target. A successful conversation should always end with a path forward. Your network grows not by talking to one person, but by having that one person introduce you to the next. The single most powerful question you can ask at the end of an informational interview is: "Thank you so much for your time. This has been incredibly helpful. Based on our conversation, is there anyone else in your organization or your network you think it would be valuable for me to speak with?" A positive response to this question is your ultimate measure of success. It means you have not only gathered intelligence but have also successfully turned a "cold" contact into a warm ally who is now actively helping you navigate the organization.

How to Maintain and Nurture Your Network

You've done the hard work. You've made the connections. You have a list of recruiters, engineers, and alumni in your professional orbit. The single biggest mistake you can make now is to let those connections go cold. A network is not a static object you acquire; it is a living garden that requires periodic, gentle cultivation. The most successful professionals understand that the real value of a network is not in the transaction of getting a job today, but in the long-term relationships that will provide advice, insight, and opportunity for decades to come.

The goal of network maintenance is not to keep a list of people you can ask for favors. It is to transition from being a "student seeking help" to being a "valued professional colleague." This requires a subtle but profound shift in mindset. You must adopt the principle of "Give Before You Get." Your aim in maintaining your network is to stay on people's radar in a positive, value-added, and non-demanding way. It is about building a community, not just a list of contacts.

Your Personal "CRM"

To do this professionally, you need a system. Do not rely on your memory. Create a simple tool, like a private spreadsheet, to act as a personal "Customer Relationship Manager" (CRM). The columns should be: Name, Title, Company, How We Met, Last Contact Date, and a Notes section. In the Notes section, capture a key, personal detail from your conversation: "Loves to ski," "Daughter is also studying engineering," "Worked on the GPS program." This simple tool is the foundation for all meaningful, long-term follow-up.

Actionable Tactic 1: The "Congratulations" Note

This is the simplest and most powerful tactic in your toolkit. LinkedIn will be your intelligence source. When you see a key contact in your network post about a promotion, a new job, or a major project success, send them a brief, genuine message of congratulations. It takes 10 seconds, costs you nothing, and reminds them of your existence in a positive, supportive way that requires no response.

The Script (via LinkedIn Message):
> "Hi Mr. Smith, I just saw your post about your promotion to Senior Principal Engineer. That's fantastic news. Congratulations!"

Actionable Tactic 2: The "Interesting Article" Share

This is a slightly more advanced tactic that demonstrates you are actively engaged in the industry and that you were listening during your conversation. If you read a fascinating article (e.g., in Aviation Week or Defense News) that is directly relevant to a contact's specific field of work (which you should have recorded in your CRM), share it with them.

The Script (via LinkedIn Message or Email):
> Subject: Interesting article on hypersonic materials
>
> "Hi Dr. Davis, I hope you're doing well. I remembered our conversation about the thermal challenges in hypersonic systems, and I came across this interesting article in Aviation Week today that I thought you might find relevant. It reminded me of the great insights you shared. Best, [Your Name]"

Why it Works: This is a "give," not a "get." You are providing them with a piece of relevant information. It demonstrates that you were listening, that you share their interests, and that you are thinking at a professional level.

Actionable Tactic 3: The "Six-Month Professional Check-in"

This is the key to keeping your most important relationships warm, especially with former internship managers and mentors. Set a recurring calendar reminder for every six months. The goal is not to ask for anything. The goal is to provide a brief, professional update on your own progress, which shows respect for the time they invested in you.

The Script (via Email):
> Subject: Quick Update from [Your Name]
>
> "Dear Ms. Evans, I hope this email finds you well. I was just thinking about my time at [Company Name] last summer and how much I learned from you and the team. I wanted to send a quick update on my progress this year.

My senior design project is going well; I'm leading the software sub-team for our autonomous rover, and we just had a successful preliminary design review. I'm also taking a graduate-level course on AI and machine learning that I'm really enjoying.

Thank you again for the incredible internship experience. The lessons I learned about [mention a specific lesson] have been invaluable.

Best regards,

[Your Name]"

Why it Works: This is a powerful, professional, and non-demanding way to stay on their radar. It shows that you are continuing to grow and succeed, which reflects positively on them as a mentor. When it comes time for full-time hiring, you are not a forgotten memory; you are the impressive young professional who has been keeping them in the loop.

From Connection to Career

We have now deconstructed the often-intimidating world of networking, transforming it from a vague social art into a structured, engineering-driven process. You have a playbook. You know how to conduct reconnaissance for a career fair, how to craft a compelling elevator pitch, the verbatim scripts to confidently reach out to alumni, and the strategic framework to turn a summer internship into a powerful network-building machine. You have all the tools.

But the tools are only as effective as the mindset that wields them. It is crucial to remember the core principle we began with: networking is not a short-term transaction; it is a long-term investment in professional relationships. Your goal in any of these interactions is not to "get a job." Your goal is to learn, to connect, and to build a reputation as a curious, respectful, and competent engineer. If you approach every conversation with a genuine desire to understand the other person's world, the opportunities will follow as a natural consequence.

The most successful networkers are not the smoothest talkers; they are the best listeners. They are the ones who are insatiably curious about the challenges others are facing and the paths they have traveled. This authenticity is your greatest asset. It is what will make a senior engineer want to spend their valuable time mentoring you. It is what will make an alumnus forward your resume to a hiring manager. They are not just helping a random student; they are helping a promising future colleague in whom they see a spark of their own passion.

This is a long game. The connections you make today, whether it is the recruiter you impress at the career fair, the alumnus who gives you 15 minutes of their time, or the senior engineer you have coffee with during your internship, are the seeds of your future career. Nurture them. Follow up. Send a thank you note. Send a LinkedIn message six months later to share a success on your capstone project. By investing in these relationships with professionalism and gratitude, you are building more than just a list of contacts; you are building a professional support system that will serve you for decades to come. You are transforming yourself from a faceless applicant into a known, respected, and sought-after member of the professional community before you even graduate.

Chapter 21

Acing the Interview

Congratulations. If you've reached this stage, it means your resume, your cover letter, and your networking have done their job. The company has looked at your credentials on paper and concluded that you are a qualified candidate. Now, they want to meet the person behind the paper. This is the moment of transition, where the abstract data points of your application must coalesce into a competent, confident, and professional human being sitting across the table from them.

The interview is the next filter. It's where the company assesses not just your technical knowledge, but your thought process, your communication skills, your personality, and your potential to be a good teammate. For many students, this is the most nerve-wracking part of the entire process, a high-stakes oral exam where they feel every word is being judged. But it doesn't have to be this way. With the right preparation and the right mindset, the interview can be transformed from a stressful interrogation into what it is truly meant to be: a professional conversation between future colleagues.

Success in an engineering interview isn't about having all the right answers memorized. It's about having a structured way to approach any question that comes your way. It is about demonstrating not just what you know, but *how you think*. This chapter will provide you with the tools, the frameworks, and the insider's perspective to confidently navigate both the behavioral and technical portions of a defense industry interview. Our goal is to replace your anxiety with a concrete, actionable plan.

48 Hours Before the Interview

Your performance in the interview is not determined in the interview itself; it is determined by the quality of your preparation in the 48 hours leading up to it. "Winging it" is not an option. Walking into a technical interview unprepared is the professional equivalent of walking into a final exam without ever attending a lecture. It is a sign of disrespect for the interviewer's time, and it is the fastest way to signal that you are not a serious candidate. The engineers sitting across the table from you are taking hours out of their incredibly busy schedules, away from critical program deadlines, to speak with you. You owe them the courtesy of being meticulously prepared.

This preparation is not about memorizing answers. That approach is fragile and often leads to robotic, uninspired responses. Instead, this is about building a confident command of your own story and your own knowledge. It is a structured process of reconnaissance, rehearsal, and logistical planning that will arm you with the information and the confidence you need to succeed. The goal is to walk into that room so well-prepared, so fluent in your own experiences, and so knowledgeable about their company that you are no longer nervous, but excited for the opportunity to have a professional, intelligent conversation.

First, you must conduct thorough recon and intelligence gathering. Go back to the job description and your Keyword Blueprint. Re-read it carefully. Go to the company's website and read the "About Us" and "Products" sections for the specific division you are interviewing with. Do a quick news search for any major, recent contract wins. Knowing these details is how you will craft your specific, compelling answers and questions.

Next, identify your weaknesses and prepare your path forward. No candidate is perfect, and a good interviewer will often try to probe for your knowledge gaps. Self-awareness is a sign of professional maturity. Before the interview, perform an honest self-assessment against the job description. Do they list a piece of software you've only used once? Is there a technical concept you're rusty on? Acknowledging these gaps is step one. Step two is formulating a plan to address them. This is a critical skill that will come up again and again in performance reviews once you are hired.

Example Self-Assessment: "The job requires experience with FEA. I did well in my FEA class, but my capstone project was more focused on control systems."

Example Prepared Answer: If asked about your FEA skills, you can say: "My primary hands-on experience has been with control systems, which is a strength I can bring to the team. While my direct FEA experience is from my coursework, I recognize it's a key skill for this role, and I've already started an online tutorial series on ANSYS to get my skills up to speed. I'm a fast learner and I'm confident I could become proficient quickly."

Your most powerful tool for organizing your thoughts is the Interview 'Mission Prep' Sheet located in Appendix B. Before every interview, print out a fresh copy. The physical act of writing down your research on the company, outlining your 3-5 strongest STAR stories, and preparing your intelligent questions to ask them will transform your preparation from a passive activity (worrying) into an active one (strategizing). We will also include a 'Post-Mission' section for self-debrief to ensure you learn and improve from every experience.

Logistics is a key detail that must not be overlooked. For a virtual interview, test your camera and microphone, ensure you have a professional, clean background, and dress in full professional attire. For an in-person interview, print multiple copies of your resume and plan your route to arrive 10-15 minutes early.

Long-Term Preparation

While the 48-hour "Mission Prep" is your crucial final cram session, the truly elite candidates begin their preparation months, or even years, in advance. An interview is a performance, and like any high-stakes performance, it requires dedicated practice long before you ever step onto the stage. The confidence you project in an interview, the ability to articulate your thoughts clearly under pressure, is not something you can fake. It is an earned skill, a muscle that must be developed over time through deliberate practice.

Your university provides a wealth of resources to help you build this muscle; do not let them go to waste. The students who feel most comfortable in interviews are the ones who have taken the time to practice in low-stakes environments, who have sought out feedback, and who have made their mistakes in a practice room instead of a conference room. This long-term preparation is what separates the merely qualified from the truly impressive.

Your university's career center is one of your most valuable training grounds. They almost certainly offer mock interviews, where you can sit down with a trained professional and practice answering common interview questions in a realistic environment. This is an absolutely essential experience. You will get direct, honest feedback on your answers, your body language, and your overall professional presence. It is far better to make your rookie mistakes and receive constructive criticism in a practice session than in a real interview where a job is on the line. Make it a goal to do at least one mock interview each academic year.

Don't limit your practice to the career center, however. Find a trusted, ambitious peer, perhaps someone from your senior design team, and practice with each other. Take turns asking each other the common behavioral and technical questions in this chapter. This informal practice helps you build fluency and confidence. Furthermore, the alumni or industry professionals you connect with during your informational interviews are an incredible resource. After a good conversation, you can ask, "As someone who interviews candidates, what is the single biggest mistake you see young engineers make?" Their answers will be pure gold.

Many engineers are brilliant thinkers but struggle to articulate their ideas clearly and confidently. Any opportunity to practice public speaking is a direct investment in your future interview performance. Join a club like Toastmasters. Volunteer to be the presenter for your team's project updates. Take a public speaking elective. The ability to stand in front of a room of senior engineers and clearly and concisely explain a complex technical topic is a superpower, both in an interview and in your future career. Do not shy away from these opportunities; seek them out.

The Two Types of Interview Questions

An engineering interview is a structured assessment designed to answer two fundamental questions: "Can you do the job?" and "Will you be a good fit for our team?" To answer these, interviewers use two distinct but complementary types of questions. Your ability to recognize the intent behind each type of question and to respond with the right kind of answer is the key to mastering the interview.

1. Behavioral Questions

Behavioral questions are the backbone of most modern interviews. They are not hypothetically asking what you *would* do; they are asking you to provide evidence of what you *have done*. The guiding philosophy is that past performance is the best predictor of future performance. When an interviewer asks, "Tell me about a time you faced a challenge," they are not just looking for a good story. They are trying to assess your problem-solving process, your resilience, your communication skills, and your professional maturity by examining a real-world example from your past.

These questions are your opportunity to bring your resume to life. They are invitations to tell the detailed stories behind the bullet points you so carefully crafted. Your tool for answering these questions is the STAR method, which we will deconstruct in the next section. Your preparation involves anticipating these questions and having a library of your best, most relevant stories ready to go, polished and practiced.

<u>*Common General Behavioral Questions:*</u>

"Tell me about a time you had to work on a team to solve a difficult problem."

"Give me an example of a time you had to learn a new skill or technology quickly."

"Describe a significant project you worked on. What was your specific role?"

"Tell me about a time a project or a task didn't go as planned. What happened, and what did you learn from it?"

Behavioral Questions Tailored for the Defense Sector:

"Tell me about a time you had to work with a very strict set of requirements or a formal process. How did you handle it?"

"Describe a project where attention to detail was absolutely critical. How did you ensure the quality of your work?"

"Give me an example of a time you had to negotiate conflicting requirements or priorities on a project."

2. Technical Questions

Technical questions are designed to test your core engineering knowledge and your thought process. They are not "gotcha" questions. The interviewer is not trying to make you fail; they are trying to understand how you think when faced with a real engineering problem.

The Live Demonstration

It is increasingly common for interviews to include a practical, hands-on demonstration of your skills. The goal is not just to see your proficiency, but to see *how you work* and *how you think* under pressure.

For Mechanical Engineers: You may be given a simple design problem and asked to create a basic model in SOLIDWORKS or Creo live on a shared screen.

For Software Engineers: You will almost certainly be given a simple coding problem and asked to solve it in a shared text editor.

For all disciplines: The key is the same: think out loud. Narrate your process.

Discipline-Specific Technical Questions (Examples):

For Mechanical & Aerospace Engineers: "Looking at this physical object, what can you tell me about how it was likely manufactured?" "Tell me about the last time you worked on a car. What was the problem and how did you fix it?"

For Electrical & Computer Engineers: "You're at a test bench and you see this noisy signal on an oscilloscope. What are the first three things you would check?" "Tell me about a time you had to troubleshoot a circuit you built."

For Chemical Engineers: In an interview, you might be asked to explain the best way to manage the thermal decomposition of a specific material. You may be tasked with recommending a specific epoxy or adhesive to bond a composite structure on a satellite.

Applying the STAR Method in Your Interview

In Chapter 17, we introduced the STAR method as the framework for engineering powerful, concise, and quantifiable bullet points for your resume. On your resume, a STAR statement is a dense, one-line summary of an accomplishment. In an interview, your task is to take that same accomplishment and expand it into a compelling, detailed, and conversational story. This is the most important skill in a behavioral interview.

While the framework is the same, the execution is different. You are moving from a written, data-dense format to a verbal, narrative format. The goal is no longer just to state the facts, but to bring the experience to life, showcasing your thought process, your problem-solving skills, and your professional maturity along the way.

Deconstructing the STAR Method for an Interview:

S - Situation: Briefly describe the context. Set the scene in one or two clear, concise sentences.

T - Task: Describe your specific responsibility or goal in that situation.

A - Action: This is the heart of your story and should be the longest part of your answer. Describe the specific, concrete steps *you* took. Use "I" statements.

R - Result: Conclude by explaining the quantifiable outcome of your actions.

Putting It All Together for a Gold-Standard STAR Story

The Interviewer's Question: "Tell me about a time you faced a difficult technical challenge on a project."

Your "Gold-Standard" Answer:

(Situation): "On my senior capstone project, our autonomous rover was required by our 'System Requirements Document' to survive a 3-foot drop test. During our initial design phase, the FEA simulations I ran showed that our primary chassis bracket would fail, with the stress exceeding the material's yield strength by nearly 50%."

(Task): "As the lead for the mechanical sub-team, my specific task was to redesign that bracket to meet a 2.5 factor of safety

requirement, but with the added critical constraint from our program manager (our professor) of not increasing the total chassis weight by more than 5%, to preserve our battery life."

(Action): "I approached this by conducting a formal trade study. First, I analyzed the existing design to understand the primary load paths. Then, I developed and modeled three potential solutions in SOLIDWORKS: a simple 'brute force' option using thicker aluminum, a high-cost option using the same geometry but with titanium, and finally, a topology-optimized design that added strategic ribbing while removing material in low-stress areas. I used ANSYS to run a structural analysis on all three, and I presented the results to my team not just in terms of performance but also based on the cost from our Bill of Materials (BOM) and the manufacturability, which I had already discussed with our university's machinist."

(Result): "The data was clear. The topology-optimized design was the optimal solution. It successfully achieved a 2.7 factor of safety in the final validation analysis, actually reducing the part's weight by 10% and keeping us on schedule and under budget. We presented this trade study at our Preliminary Design Review (PDR), and it was the core evidence that proved our design was sound. Most importantly, the physical rover survived three drop tests during validation with no damage."

This answer is a case study in being a professional defense engineer. It uses the language of the industry (FEA, factor of safety, trade study, PDR, BOM, requirements). It demonstrates a methodical, data-driven process, and it shows you can balance performance with cost and schedule constraints.

Why This is So Effective

By following this structured, four-part narrative, you are doing far more than just answering a question. You are giving the interviewer a complete, self-contained case study of you being an effective engineer. It is a powerful psychological tool that demonstrates a whole suite of desirable professional traits without you ever having to say, "I am a good problem-solver." You are showing them, not just telling them.

Your story proves that you can analyze a situation, take ownership of a specific task, develop and execute a logical plan of action, and achieve a positive, quantifiable result. It also demonstrates your technical competence, your commitment to a methodical process, and, most importantly, your professional maturity. This is what separates a student's answer from a professional's answer.

Navigating Technical Questions

Technical questions are designed to test your thought process. While the interviewer certainly wants a correct answer, they are often far more interested in *how* you approach a problem you've never seen before. This is a direct simulation of how you will perform as a junior engineer on their team. Your goal is to showcase a methodical, calm, and logical problem-solving process.

The Golden Rule: Think Out Loud

The single biggest mistake a candidate can make when faced with a tough technical question is to go completely silent. The interviewer has no idea what you are thinking. They cannot see your brilliant internal analysis. You must narrate your thought process from beginning to end. Verbalizing your approach does three powerful things: it shows the interviewer *how you think*, it allows them to correct a flawed assumption and guide you in the right direction, and it transforms a stressful interrogation into a collaborative, problem-solving discussion between two engineers, which is, after all, the entire point of the job.

A Playbook for the Technical Question:

Listen and Clarify: Listen to the entire question without interrupting. Then, repeat the core of the problem back in your own words. This is not just stalling for time; it is a professional habit that ensures you are solving the right problem. "Okay, so just to be clear, you're asking me to estimate the cooling requirements for this sealed electronics box with a processor inside?"

Identify and State Your Assumptions: No real-world problem comes with all the data. You must make assumptions. State them clearly and upfront. "This is a great question. To start, I'm going to assume this is a sealed aluminum box, and for a first-order analysis, I'll assume natural convection and radiation are the only modes of heat transfer to the outside air, which I'll assume is at standard room temperature, about 25°C."

Whiteboard Your Approach (Literally or Verbally): Before you jump into calculations, outline your plan. "My overall approach will be to first calculate the total heat generated inside the box. Then I'll determine the thermal resistance of the box itself via conduction. Finally, I'll calculate the heat dissipated to the environment through both convection and radiation to find the steady-state surface temperature and see if it's within a reasonable limit."

Execute and Narrate: Walk through the steps of your plan, narrating your thought process as you go. "Okay, first step, heat in. If the processor is 100 watts and the power supply is 80% efficient, that means it's also generating 25 watts of heat, so our total Q-dot is 125 watts..."

Sanity Check Your Answer: Once you arrive at an answer, gut-check it. Does it make sense? A surface temperature of 5,000 degrees is obviously wrong. Showing that you have an intuitive feel for what is reasonable is a huge sign of maturity. "The calculation gives me a surface temperature of 85°C. That's hot to the touch, but likely within the operational limits for the electronics. It feels reasonable for a first-order analysis. My next step would be to recommend a more detailed FEA model and perhaps investigate adding a heat sink or a small fan to increase our margin."

This structured, vocal approach proves something far more important than whether you memorized the right formula. It proves that you can be trusted to tackle a novel problem with rigor, logic, and professionalism.

Handling Tricky Behavioral Questions

Certain questions are designed not just to test your experience, but your character, self-awareness, and professional maturity. Generic, rehearsed answers are transparent and ineffective. Your goal is to be honest, strategic, and poised.

"What is your greatest weakness?"

Aim to demonstrate self-awareness and a commitment to improvement. Never give a cliché "humblebrag" like "I'm a perfectionist" or "I work too hard."

The Framework: Name a real but manageable weakness, explain the context, and (most importantly) describe the specific, concrete actions you are taking to improve it.

Good Answer: "In the past, I've sometimes been hesitant to ask for help on a difficult problem, as I really value being self-reliant. I learned from my internship experience that it's much more efficient to struggle for a defined period and then seek advice from a senior engineer. It's something I'm actively working on improving by setting a 'one-hour rule' for myself before I ask for guidance, and it's a process that has made me a much more effective problem-solver."

"Tell me about a time you had a conflict with a teammate."

The interviewer is testing your emotional intelligence and your ability to function professionally. Do not blame the other person. Focus on the process of resolving the disagreement.

The Framework: Describe a professional disagreement (not a personal drama). Focus on a technical issue. Explain how you listened to their perspective, how you used data and objective evidence to make your case, and how you worked toward a compromise that was best for the project.

Good Answer: "On my capstone team, my software lead and I had a disagreement about which sensor to use for obstacle detection. He favored a simple, inexpensive sonar, while I was concerned about its accuracy. Rather than argue, I proposed a quick trade study. I spent an afternoon researching both options and put together a one-page comparison matrix showing that while the LIDAR sensor cost $50 more, it offered a 40% improvement in measurement resolution and would save us weeks of software development time in writing filtering algorithms. After seeing the data, we both agreed the LIDAR was the right long-term choice for the project."

"Why do you want to work in the defense industry?"

They are looking for a mature, mission-focused answer. Seize this chance to show you've absorbed the core themes of this book.

The Framework: Combine the technical challenge with the mission.

Good Answer: "I'm drawn to the unique and incredibly challenging technical problems in this industry. I'm motivated by the opportunity to work on high-consequence systems where rigor, process, and attention to detail are paramount. Above all, I find the mission of contributing to national security, and building the tools that protect our service members, to be very compelling."

From Candidate to Colleague

The engineering interview, at its heart, is a transition. It is the moment you stop being a student on paper and start being a potential colleague in person. It is not a test of your memory, but a demonstration of your mind. Your goal is not to have every answer, but to have a structured, professional, and confident approach to every question.

By now, you have all the tools you need. You have a long-term strategy for building your communication skills through mock interviews and public speaking. You have a short-term, 48-hour mission prep plan to conduct reconnaissance on the role and prepare your specific talking points. You understand the difference between a behavioral question, which demands a story, and a technical question, which demands a process.

You have mastered the STAR method, the ultimate framework for transforming your resume's bullet points into compelling narratives. You know the Golden Rule of technical questions: to "think out loud." And you have a strategic, honest approach to handling the tricky questions that are designed to test your self-awareness and your motivation.

Ultimately, the engineers sitting across the table from you are looking for two things: competence and character. Your technical answers and your project stories will prove your competence. The way you communicate, the way you handle a problem you don't immediately know the answer to, and the thoughtful questions you ask will prove your character. Walk into that room prepared, be confident in the skills you have built, and be genuinely curious about the challenges they are trying to solve. If you do, you will walk out not just as a candidate, but as their next great hire.

Chapter 22

The Security Clearance, Demystified

You've aced the interview. You've received a contingent job offer. You have proven your technical competence and your professional potential. Congratulations. You have passed the company's filter. Now, you must pass the government's. For the vast majority of jobs in the defense industry, your employment is contingent upon your ability to obtain and maintain a security clearance.

This process is, without question, the single biggest source of anxiety, mystery, and misinformation for aspiring defense engineers. It is a deep and intensely personal background check conducted by the U.S. government to determine if you can be trusted with the nation's secrets. The very idea of having your entire life investigated can be incredibly intimidating, conjuring up images of stern-faced investigators in dark suits and stressful interrogations where a single wrong answer could derail your entire career.

This chapter is your demystification guide. We will pull back the curtain on this entire process. We will walk through what a clearance is, what the government is *really* looking for (and what they are not), how to prepare for and fill out the single most important form of your life (the SF-86), and what to expect during the investigation itself. The process is rigorous, but it is not designed to trick you, to judge you, or to punish you for past mistakes. It is a structured, professional, and remarkably consistent process designed to assess one thing and one thing only: your trustworthiness. For the vast majority of responsible U.S. citizens, it is a straightforward, if lengthy, journey.

A Brief History of Trust

Before we dive into the details of the modern process, it's important to understand *why* security clearances exist. The concept of a formal, standardized background investigation for government and industry personnel is a relatively modern invention, born out of the immense national security challenges of the 20th century.

The story begins in earnest with the Manhattan Project during World War II. The United States had gathered the world's top scientists to build the atomic bomb in secret locations like Los Alamos, New Mexico, and Oak Ridge, Tennessee. The government realized that the project's secrecy was non-negotiable; if the enemy discovered the program or, worse, stole its secrets, the consequences could be catastrophic. This led to the creation of the first large-scale, formalized personnel security program. The men and women working on the project were subjected to deep background checks to ensure their loyalty and trustworthiness.

After the war, the onset of the Cold War and the revelation of several high-profile espionage cases made it clear that a permanent, government-wide system was needed. The Atomic Energy Act of 1946 established the clearance system for the new Atomic Energy Commission (the precursor to the Department of Energy - DOE), creating the "Q" and "L" clearances to protect the nation's nuclear secrets. A few years later, a parallel system was established for the Department of Defense (DoD) and other agencies to protect military and intelligence secrets. This historical divide is why the DOE and DoD still maintain their own, largely reciprocal, clearance systems today. Their foundational missions, one centered on nuclear secrets and the other on military secrets were born in different legislative acts.

The modern process, with its 13 Adjudicative Guidelines and the "Whole Person" concept, is the direct result of decades of lessons learned during this high-stakes competition. It is a system designed to be as fair and objective as possible, a structured response to the very real threat of espionage.

What is a Security Clearance?

A security clearance is a formal determination by the U.S. government that an individual is eligible for access to classified national security information. It is not a rank or a permanent award; it is a status of trust that is granted based on a thorough investigation and must be maintained over time. There are three main levels you will encounter:

Confidential: This is the lowest level, for information whose unauthorized disclosure could cause "damage" to national security.

Secret: This is the most common level required for most engineers. It involves a more detailed investigation. Unauthorized disclosure could cause "serious damage" to national security.

Top Secret (TS): This is the highest level, required for work on the most sensitive programs. It involves a much more in-depth, comprehensive investigation. Unauthorized disclosure could cause "exceptionally grave damage" to national security.

Beyond the Three Levels: SCI, SAPs, DOE Clearances, and Polygraphs

While Confidential, Secret, and Top Secret are the foundational "clearance levels," they are not the end of the story. For engineers working on the most sensitive and advanced programs, there are additional layers of access and different types of clearances that are important to understand.

SCI (Sensitive Compartmented Information): You will often see a job posting that requires a "TS/SCI" clearance. SCI is not a higher clearance level than Top Secret; it is a type of access granted for specific, highly sensitive intelligence programs. Think of it this way: your Top Secret clearance is like having a key to the main library. An SCI "ticket" or "caveat" is a special key that lets you into a private, restricted room in the back of that library. Access is granted on a strict "need-to-know" basis, and you will only be "read into" the specific SCI compartments that are relevant to your project. Additionally, eligibility for SCI often requires a separate, more focused investigation in addition to the primary clearance investigation, a process that delves even deeper into your background to ensure the highest level of trust. This system is designed to protect the nation's most sensitive intelligence sources and methods.

SAPs (Special Access Programs): This is the next level of compartmentalization, often referred to colloquially as "black programs." A SAP is a highly classified program with security measures and access restrictions that exceed even those of standard SCI information. These programs are established to protect the nation's most advanced and groundbreaking technologies, things like a new stealth aircraft or a next-generation satellite system. Being granted access to a SAP is a significant step that requires an even more rigorous vetting process and a deep level of trust.

The Department of Energy (DOE) "Q" and "L" Clearances: This is a parallel but equivalent system that runs out of the Department of Energy. The DOE is responsible for the nation's nuclear weapons stockpile, nuclear reactors, and other highly sensitive nuclear information. As such, they have their own clearance system.

An "L" Clearance is generally equivalent to a DoD Secret clearance.

A "Q" Clearance is generally equivalent to a DoD Top Secret clearance.

These clearances are often reciprocal, meaning that if you hold a DoD Top Secret clearance, the process of granting you an equivalent DOE Q clearance is often much faster, and vice versa. Engineers working at the NNSA National Laboratories (like Sandia, Los Alamos, or Lawrence Livermore) will hold DOE Q clearances.

The Polygraph Examination: For certain positions requiring the highest levels of trust a polygraph examination may be required as part of the investigative process. These programs are typically those involving access to specific SCI compartments, certain SAPs, or work with some intelligence agencies. It is important to understand what this is and what it is not. A polygraph is an administrative tool used to aid in the verification of the information you provided on your SF-86 and during your interviews. There are two primary types you might encounter: a Counterintelligence Polygraph and a Lifestyle Polygraph. Crucially, you cannot and should not attempt to "prepare" for a polygraph. The only instruction you will ever be given, and the only one we can provide here, is to be completely and consistently truthful throughout the entire security clearance process. Any attempt to learn or use countermeasures is illegal and will result in the immediate and permanent denial of your clearance.

The Most Important Rule: You cannot apply for a security clearance on your own. A defense contractor (or a government agency) must sponsor you for one after they have given you a job offer. The company pays for the investigation, which is conducted by the government's Defense Counterintelligence and Security Agency (DCSA).

Understanding the "Interim" Clearance

It's important to understand that you will not have to wait for the full 6-18 month investigation to be completed before you can start doing meaningful work. For most candidates, the first step is receiving an "interim" clearance.

What It Is: An interim clearance (sometimes called "interim eligibility") is a temporary clearance that may be granted after the initial, favorable review of your SF-86 submission and a check of the major national databases. It is a preliminary decision, made in a matter of a few weeks or months, that allows you to begin working on and handling classified material up to the level you are being investigated for (e.g., an interim Secret allows you to handle Secret material).

Why It Matters: The interim clearance is the key that unlocks your ability to join your team and start contributing. It is the moment you transition from onboarding and unclassified training to being "read into" your program and entering the "vault." Your first 90 days on the job will almost certainly be conducted under an interim clearance while the full investigation continues in the background.

The Caveat: An interim clearance is not guaranteed and can be withdrawn if unfavorable information is discovered during the full investigation. This is another powerful reason why complete honesty on the SF-86 is so critical.

The "Whole Person" Concept

Before we dive into the details, you must understand the guiding philosophy of the entire security clearance process. It is called the "Whole

Person" Concept. The investigators and adjudicators who review your case are not trying to "fail" you. They are not looking for perfect people, because perfect people do not exist. They are looking for trustworthy people. Trustworthiness, defined by honesty and reliability, is the single most important attribute.

This means they will consider your entire life, both the good and the bad, to assess your overall character, reliability, and judgment. They are assessing you against a formal set of 13 Adjudicative Guidelines, which cover everything from allegiance to the U.S. to financial responsibility and personal conduct. A single mistake you made as a teenager is highly unlikely to disqualify you, as long as you are honest about it and it does not represent a pattern of behavior. A *pattern* of poor judgment or, most critically, a *failure to be honest* about your past, however, is a major problem.

The government's primary concerns are simple: Are you loyal to the United States, and are you vulnerable? Is there anything in your past like a hidden crime, a mountain of secret debt, a substance abuse problem that a foreign adversary could use to blackmail or coerce you into revealing classified information? The entire process is a risk assessment based on that question. Honesty is your greatest asset because it proves you are not vulnerable to blackmail.

The SF-86: Your Life on Paper

The entire process begins with the *Standard Form 86 (SF-86)*, *"Questionnaire for National Security Positions."* After you accept your contingent job offer, you will be sent a link to an online government portal called e-QIP to fill out this form.

The SF-86 is one of the most comprehensive documents you will ever complete. It is a lengthy, minutely detailed account of your life, typically looking back over the last 7-10 years. It will ask for:

- Everywhere you have lived, with dates and a person who knew you at that address.
- Everywhere you have worked, with dates, your supervisor's name, and their contact information.
- All of your education after high school.
- Detailed information about your relatives, including their citizenship.

- A list of your foreign contacts and every country you have ever traveled to.
- A detailed look at your financial history, including any bankruptcies, delinquencies, or collections.
- Your police record, including any and all arrests or charges, even if they were minor, and even if they were dismissed or expunged.
- A history of any drug or alcohol-related issues.

How to Prepare for the SF-86

The SF-86 can be incredibly stressful if you are unprepared. The best way to reduce this stress is to start gathering the required information long before you ever get the link to the form.

Create a "Life Spreadsheet" Today: Open a spreadsheet and start a new tab for each category (Residences, Employment, etc.). Begin filling it out now, while the information is fresh. It is much easier to find your old supervisor's phone number today than it will be two years from now under a tight deadline.

Pull Your Credit Report: You are entitled to a free report every year from the three major bureaus. Get it. You need to know exactly what is on your financial record, so you are not surprised by a forgotten collection account from an old utility bill.

Gather Your Documents: Find your passport, birth certificate, and any other vital documents you will need to reference.

The Golden Rule for the SF-86: Complete and Utter Honesty

This is the most critical advice in this entire book. You must be 100% truthful on the SF-86. The investigators have an incredible ability to verify the information you provide. They will pull your credit report, check national and local court records, and interview your friends, neighbors, and former employers. The single fastest way to be denied a clearance is to be caught intentionally hiding something or lying on the form. This is called falsification, and it is a federal offense.

An adjudicator can mitigate almost any issue if you are honest about it. They cannot mitigate a lie.

Did you try marijuana a few times in college in a state where it was legal? Disclose it. It is very often a mitigatable issue, especially if it was not recent. Lying about it and getting caught is not.

Did you get arrested for underage drinking at 19? Disclose it. A minor incident from your youth that shows no pattern is understandable. Hiding it is a sign of a character that cannot be trusted today.

Common Myths vs. Realities: Alleviating Your Concerns

This is where most of the anxiety comes from: misconceptions and rumors. Let's replace those myths with reassuring facts.

The Myth	*The Reality*
"Any past drug use is an automatic disqualifier."	False. Past, limited usage that you are honest about is often a mitigatable issue. *Honesty is the key.*
"My student loan debt will prevent me from getting a clearance."	False. Normal, managed debt is not a problem. Investigators are looking for signs of financial *irresponsibility.*
"Investigators will interrogate my friends and family."	False. The process is professional, not adversarial. It is a verification process, not an interrogation.
"I have to be a perfect, boring person."	False. Having traveled extensively or having foreign friends is not a disqualifier. You simply need to *be honest.*
"Any past financial trouble is a disqualifier."	False. Financial trouble that you have been honest about and are actively working to resolve (e.g., through a payment plan) is often a mitigating factor. They are looking for financial *irresponsibility*, not the simple fact of having debt or past issues.

The Investigation and Adjudication

After you submit your SF-86, an investigator will be assigned your case. They will conduct national database checks (FBI, credit reports, etc.). For a Secret clearance, they will likely conduct a subject interview with you, either over the phone or in person. This is not an interrogation. It is a professional interview where they will walk through your SF-86 with you, clarify any confusing information, and ask follow-up questions. They will also interview your listed references.

All the information is compiled into a report. A trained government adjudicator, who you will never meet, reviews the entire file. They apply the "Whole Person Concept" and the 13 Adjudicative Guidelines and make the final decision to grant or deny the clearance.

This process takes time. Be patient. Be honest. For the vast majority of candidates, if you are a responsible U.S. citizen without a pattern of criminal activity, serious financial irresponsibility, or divided loyalties, you will be successful.

Life with a Clearance: Your Ongoing Responsibilities

Receiving your security clearance is not the end of the process; it is the beginning of a lifelong commitment to maintaining that trust. Holding a clearance is a privilege, and it comes with a set of ongoing responsibilities that will become a normal part of your professional life. Understanding these expectations from the outset is a key part of being a responsible cleared professional.

The Concept of "Reportable" Information: As a cleared individual, you are required to report certain activities and life events to your company's security officer. This is not meant to be intrusive; it is part of the partnership to mitigate potential security risks. Common reportable events include:

Foreign Travel: You will typically need to report your intention to travel internationally in advance, listing your destinations and dates.

New Foreign Contacts: If you develop a close and continuing relationship with a foreign national (e.g., a serious romantic partner, a new roommate), you will need to report it.

Significant Financial Issues: Events like filing for bankruptcy, having a home go into foreclosure, or having a significant amount of debt go to collections are often reportable.

Involvement with Law Enforcement: Any arrest or criminal charge, no matter how minor, must be reported immediately.

Annual Security Training: You will be required to complete annual security training, which will refresh you on the proper procedures for handling classified information, remind you of your reporting responsibilities, and update you on the latest counterintelligence threats.

Periodic Reinvestigations (PRs): A security clearance is not granted for life. It is periodically re-evaluated to ensure you remain a trustworthy individual. While the DCSA has largely moved to a process of Continuous Vetting (CV) rather than fixed-interval PRs, reviewing PR timelines is a helpful comparison of the depth investigation that each clearance requires. A Top-Secret clearance is typically reinvestigated every 5-7 years, while a Secret clearance is reinvestigated every 10-15 years. This process usually involves updating your SF-86 to cover the period since your last investigation.

These responsibilities may seem daunting at first, but they quickly become a routine and manageable part of a career in this industry. They embody the ongoing trust that the nation has placed in you.

The security clearance process, when viewed from the outside, can seem like an intimidating and mysterious monolith. It is natural to feel anxious. However, the most important thing you can take away is this: the system is not designed to find perfect people. It is designed to find trustworthy people. The "Whole Person" concept is a genuine philosophy that guides the entire process. Adjudicators are trained to look for patterns, not for single, isolated mistakes. They are looking for honesty, reliability, and a character that is not vulnerable to coercion or blackmail. A past mistake does not disqualify you; a current attempt to hide it does..

You now have a clear understanding of this world. You know the history, the levels of clearance, and the practical steps to prepare for the SF-86. You understand the golden rule: complete and utter honesty.

Navigating this "great filter" with confidence is a marathon, not a sprint. It begins long before you ever receive the link to the SF-86, with the responsible decisions you make every day. In the next chapter, we will move from the process to the practice, taking a deep dive into the specific adjudicative guidelines and the personal and financial habits that demonstrate the trustworthiness that is the hallmark of a great engineer.

Chapter 23

Passing the Background Check

In the last chapter, we demystified the clearance process. In this chapter, we get personal. The security clearance process, at its core, is a test of your trustworthiness, and the adjudicators who review your case are looking at your life through a single, powerful lens: they are assessing risk. Their primary question is: "Is there anything in this person's life that could make them vulnerable to blackmail, coercion, or foreign influence, or that indicates a pattern of unreliable or dishonest behavior?"

This scrutiny can feel intimidating, but it is an opportunity. The very habits of mind and life that make you a low-risk candidate for a security clearance are the same habits that build the foundation for a stable and successful career and life. This is not about being perfect, because no one is perfect. It is about demonstrating reliability, good judgment, and a commitment to living a responsible and transparent life.

This chapter is a practical, no-nonsense guide to the two areas of your life that receive the most scrutiny under that lens: your personal conduct and your financial health. Our goal is not to judge your past, but to give you a clear, actionable plan to shape your future, replacing your anxiety with confidence and a clear path forward. This is your personal fitness plan for a career in the high-stakes world of national security.

Personal Conduct – The Character and Judgement Test

This area covers a wide range of your life choices, from your online presence to your personal associations. The government is not trying to judge your morals; it is trying to assess your reliability and judgment. They are looking for patterns. A single mistake is an anomaly; a series of related mistakes can be a pattern. Your job is to be honest about any anomalies and to demonstrate that they are in the past.

Drug Involvement

This is, without question, the number one source of anxiety for most young applicants, especially given the rapidly changing legal landscape of marijuana. Let's be clear: the federal government, which grants security clearances, still considers marijuana to be an illegal substance, regardless of state law. However, the way it is adjudicated has evolved significantly.

The Adjudicator's Questions:

An adjudicator is not asking, "Did this person ever use drugs?" They are asking:

Recency: When did the use occur? Use that happened years ago in your teens is viewed very differently from use that happened last month.

Frequency: Was this a one-time experimentation, or was it a regular, habitual part of your lifestyle?

Circumstances: Did the use occur in a context of immaturity (e.g., a college party), or did it involve a more serious disregard for the law (e.g., using drugs while in a position of public trust)?

Honesty: Were you completely and utterly truthful about it on your SF-86? Lying about drug use is a far more serious offense than the use itself.

Actionable Plan and Reassurance:

Cease All Use Immediately. The single most important mitigating factor for past drug use is demonstrated abstinence. The day you decide you want a career in the defense industry is the day you must commit to a drug-free lifestyle. The longer the period between your last use and your SF-86 submission, the better.

Be Prepared to Be Honest. You must disclose any and all past use. Do not try to hide it. The investigation is thorough. They will interview your friends and references, and if there is a discrepancy, it will be discovered.

Understand Mitigation. If you used marijuana a few times in college and stopped two years ago, that is a highly mitigatable issue. You can honestly say, "That was a part of my past during a period of experimentation in college. I have since matured, I understand my responsibilities, and I have not used any illegal substances since [Date]." This shows maturity and a commitment to the rules.

Alcohol Consumption

Occasional and legal alcohol use is not an issue. Investigators are not concerned with responsible drinking. They are looking for patterns that indicate poor judgment or a potential for unreliability, where alcohol becomes a contributing factor to negative life events.

The Adjudicator's Questions:

Has alcohol consumption ever led to any incidents with law enforcement (e.g., DUI, public intoxication, disorderly conduct)?

Has alcohol ever negatively impacted your work, academic performance, or personal relationships?

Is there a pattern of excessive consumption that suggests a lack of judgment?

Actionable Plan and Reassurance:

Be Responsible. This is simple life advice that is also good security advice. If you choose to drink, do so in moderation and never, ever drink and drive. A single DUI is a very serious issue, but it is not necessarily an automatic disqualifier. It will require significant mitigation, such as completing all court-mandated counseling and demonstrating a long period of responsible behavior.

Be Honest on the SF-86. The form will ask about alcohol-related incidents. Disclose everything. Court records are easily checked. Lying about a past DUI is a certain denial. Honesty, combined with demonstrated rehabilitation, is your only path forward.

Your Digital Footprint

In the 21st century, your online persona is a part of your "whole person." Assume that anything you have posted publicly on the internet is discoverable. Investigators may review your public social media profiles (Facebook, Twitter, Instagram, Reddit, etc.). They are not looking to judge your hobbies or your political opinions, but they are looking for red flags that indicate poor judgment, extremist ideologies, or a public persona that is wildly at odds with the trustworthy image required for a cleared professional.

The Adjudicator's Questions:

Does this person associate with extremist groups or advocate for the overthrow of the U.S. government?

Does this person exhibit a pattern of untrustworthy or criminal behavior online?

Is there anything in their public profile (e.g., pictures of illegal drug use) that contradicts what they stated on their SF-86?

Actionable Plan and Reassurance:

"Scrub" Your Public Profiles Now. Before you even apply for a job, go through your entire social media history. Go back to your high school posts. Remove unprofessional photos (e.g., those showing illegal drug use or excessive drinking) and delete any posts, comments, or "likes" that could be misinterpreted or show questionable judgment, especially those involving threats, bigotry, or extremist language.

Think Before You Post. From this day forward, operate under the assumption that a security investigator could one day read everything you post, whether or not you delete it. This is not about hiding who you are; it's about presenting a professional, mature, and reliable image to the world.

Financial Fitness – The Vulnerability and Reliability Test

Financial problems are one of the leading causes of clearance denial. This is not a moral judgment about wealth. The government's logic is straightforward: a person who is drowning in debt and cannot manage their money is, first, potentially more susceptible to being tempted by a bribe in exchange for classified information, and second, may be seen as unreliable in their general conduct.

Understanding "Good" Debt vs. "Bad" Debt

The issue is not having debt; it's having *delinquent* or *unmanaged* debt.

"Good" Debt: Taking out student loans to invest in your education, having a reasonable car payment, or using a credit card and paying it off every month are all normal parts of modern life and are not held against you. This shows that you are a responsible participant in the financial system.

"Bad" Debt: Investigators are looking for signs that you are not living within your means or are ignoring your financial obligations. This includes:

- *Accounts in Collections:* A bill (medical, utility, credit card) that you failed to pay for so long that the original creditor sold the debt to a collection agency.
- *A History of Late Payments:* A pattern of paying your bills 30, 60, or 90+ days late.
- *Defaulting on a Loan:* Failing to pay back a loan as agreed.
- *A recent Bankruptcy.*

Your Financial Fitness Action Plan

Get your financial house in order *now*. This is one of the most important things you can do to prepare for a career in this industry.

Get Your Credit Report: You are entitled to a free credit report every year from each of the three major bureaus (Equifax, Experian, TransUnion) via the official website, AnnualCreditReport.com. Do this today. Your credit report is the primary document the investigators will use. You need to know exactly what is on it.

Review and Dispute: Read through your credit report line by line. Is there an error? A bill from a former roommate that is incorrectly listed under your name? Dispute it immediately through the credit bureau's official process.

Pay Your Bills on Time, Every Time: This is the single most important financial habit you can build. It is a direct, quantifiable demonstration of your reliability. Set up automatic payments for all of your recurring bills.

Address Problems Head-On: If you have an account in collections, do not ignore it. This is a critical mistake. Hiding from a problem is a sign of unreliability. Instead, call the collection agency. Acknowledge the debt.

And, most importantly, set up a payment plan. Even if you can only afford to pay a small amount each month, the act of creating and adhering to a payment plan is a powerful mitigating factor. It shows the adjudicator that you are not ignoring your responsibilities and are actively working to resolve the issue. An old debt that you are responsibly paying off is a far, far smaller issue than a new debt that you are ignoring.

The Golden Rule of Mitigation

The path to a successful background check begins long before you fill out the SF-86. It is the sum of the small, responsible decisions you make every day. For almost any past mistake you might be worried about, from drug use to financial trouble, the path to putting it behind you is the same. Adjudicators look for two key things:

Time: The more time that has passed between the negative behavior and the present, the less relevant it is considered.

Honesty & Corrective Action: The willingness to be completely honest about the issue on your SF-86, combined with demonstrating the positive steps you have taken to correct it (e.g., stopping the behavior, entering a payment plan for a debt), is the most powerful mitigating factor of all.

By being mindful of your conduct and diligent with your finances now, you are not just preparing for a clearance, you are building the trustworthy character that is the hallmark of a great engineer in the defense industry."

Part 7

You're In. Now What?
(The First Five Years)

Chapter 24

Your First 90 Days

You've accepted the offer. You've cleared the background check. You have your start date and a crisp, new ID badge with your slightly-too-serious photo on it. The long, stressful, and uncertain process of getting the job is over. The mission is complete. The months, or even years, of hard work in your engineering program, the late nights studying for exams, the intense focus of your capstone project, the nerve-wracking interviews, it has all culminated in this moment. You have earned your place.

Now, a new mission begins.

The transition from the academic world, with its defined semesters, clear grading rubrics, and predictable rhythms, to the professional world of a large defense contractor is a significant and often disorienting one. The culture is different, the expectations are new, and the "unwritten rules" of this complex, high-stakes environment can be hard to figure out. The first 90 days of your employment are a critical window where you will set the tone for your entire career. The first impressions you make during this period, of your work ethic, your attitude, your professionalism, your coachability, will have a lasting and powerful impact on your reputation.

This is not an exaggeration. The senior engineers and managers you work with will form their foundational opinion of you during this time. They are assessing you, not just as a new hire, but as a potential long-term investment. Are you the kind of engineer they want on their team for the next decade? Are you a future leader? Or are you someone who will require constant supervision? The answers to these questions are formulated in these first three months.

This chapter is your detailed, week-by-week guide to navigating those crucial first 90 days. Think of it as your second, extended interview; a real-world audition where your performance will determine your future opportunities. Our goal is to transform you from a nervous "new hire" into a respected, reliable, and high-potential engineer, a trusted member of the team. We will follow the journey of a fictional new engineer, "Alex," to make this journey tangible and real.

The First Week (Days 1-7)

Let's be honest and set a realistic expectation: your first week on the job will likely be one of the most unproductive and overwhelming weeks of your professional life. You will be a ghost in the machine. You will be buried in an avalanche of paperwork from HR, filling out forms for your benefits, your direct deposit, and your tax withholdings. You will spend hours, if not days, in mandatory online training modules, clicking through slides on security protocols, ethics, intellectual property, and company policies. You will be introduced to a dozen people in a rapid-fire series of handshakes, and you will almost certainly forget their names within thirty seconds.

You will be given access to a new computer, new software, and a new, labyrinthine internal network of shared drives and collaboration tools, and you will not know how any of it works. You will struggle to find the right form, the right server, the right person. You will sit in your first team meetings, and the conversation will wash over you in a tidal wave of acronyms and program-specific jargon that sounds like a foreign language. You will hear about the problems with the "TRM for the new sensor suite" and the "risk burndown for the CDR," and you will have absolutely no idea what anyone is talking about. It is very easy to feel completely and utterly useless, to feel a sense of "imposter syndrome," to wonder if the company made a mistake in hiring you.

This feeling is completely normal. This is not a sign that you are failing; it is a rite of passage. Every single person in that room, from the senior technical fellow to the program manager, had the exact same experience on their first week. No one expects you to contribute technically in your first few days. They expect you to be a bit lost. Your primary goal is not to be a technical superstar; it is to be a professional sponge. Your job is to listen, to learn, to ask clarifying questions about the process, and to start building a mental map of your new environment.

Your Action Plan for Being a Professional Sponge

The Notebook is Your Most Important Tool: On day one, show up with a professional, physical notebook and a good pen. Not a laptop, which can create a physical and psychological barrier between you and the person you're talking to. A notebook. In every single meeting, training session, and introduction, take it out and write things down. This is your personal intelligence-gathering device.

Write down the names of your teammates and their specific roles ("Dave - Lead Structures Analyst," "Susan - Thermal SME"). Every acronym you hear (and leave a separate page in the back of your notebook to create your own personal glossary; look them up later on the company's internal network). The titles of key program documents that people mention. The names of the software tools the team uses. The step-by-step instructions for logging into a new system.

This simple act has a powerful, subconscious effect on your new colleagues. It shows that you are engaged, that you are taking the role seriously, and most importantly, that you respect their time by not having to ask the same question twice. The new hire who is always writing things down is immediately seen as more professional, organized, and diligent than the one who is just sitting there, trying to absorb everything through osmosis.

Your top priority is to build your "Who's Who" map and learn who is on your immediate team and what each person does. Don't just learn their names; learn their roles and their expertise. By the end of the first week, you should be able to sketch out a simple organizational chart of your team. Who is your direct functional manager? Who is the lead engineer for your specific project? Who are the other junior engineers? Who is the seasoned senior engineer who has been on this program for 20 years and carries the burden of tribal knowledge? Understanding the team's structure and who the go-to experts are is the first step to becoming an effective teammate.

Listen in meetings. Look at the company's internal directory. When your manager introduces you to someone, write down their name and what they do in your notebook. Don't be afraid to ask, "Sorry, I'm trying to get the lay of the land. What is your role on the team?" People appreciate the effort.

The Power of Listening
In your first few team meetings, your job is to be an active listener. Do not feel pressured to speak or to have an opinion on a technical matter. You have not yet earned the right to have a strong opinion because you do not yet have the context. Your goal is to learn that context before you try to contribute.

Listen for the Language: Absorb the acronyms and the way people talk.

Listen for the Problems: What are the current challenges and successes of the project? What is the "long pole in the tent" that everyone is worried about? This will give you a sense of the team's priorities.

Listen for the Dynamics: Get a feel for the team's culture. Is it formal or informal? How does the lead engineer interact with the junior engineers? Who are the people who seem to have the most influence, regardless of their official title?

By the end of your first week, you will still feel a bit lost. But you will have a notebook full of intelligence, a basic understanding of your team's structure, and you will have made a powerful first impression as a professional, engaged, and observant new hire. You have begun the process of earning your place.

The First Month (Days 8-30)

After the initial whirlwind of onboarding, you will be given your first real engineering task. This is the moment you have been waiting for, but it can also be a source of significant anxiety. Your first assignment will likely be a small, well-defined, and seemingly simple task. It might be updating a CAD drawing with a few "red-lined" changes from a senior engineer. It might be writing a simple MATLAB script to parse a small set of test data. It might be researching a new component and summarizing its data sheet.

It is very easy to look at this first task and feel a sense of letdown. After four years of complex, theoretical coursework, your first real engineering job is... this? Do not fall into this trap. Your first task is not about the task itself; it is a test. Your manager and your new teammates are watching. They are assessing your attention to detail, your ability to follow instructions, the quality of your work, and your problem-solving process. Your performance on this first, simple task is what will determine if you can be trusted with a second, more complex task. Flawless execution on the "simple" things is the fastest way to earn the right to work on the "hard" things.

Your Action Plan for Flawless Execution

Deconstruct the Task and Ask Clarifying Questions Up Front: Before you dive in, you must be absolutely certain that you understand the "definition of done." The worst thing you can do is spend a week working on something only to find out you misunderstood the assignment. After your manager or mentor gives you the task, take a moment to process it, and then repeat it back to them in your own words to ensure you are aligned.

Verbatim Script:

> *"Okay, just so I'm clear, you want me to take the red-lined markups on this drawing, incorporate them into the SOLIDWORKS model, and then generate a revised, two-page PDF drawing that complies with the company's standard drafting practices. The desired output is the PDF, and you'd like me to have it done by Thursday. Is that correct?"*

This simple act of paraphrasing the instructions shows a level of professional diligence that is incredibly rare in a new hire, and it will save you a world of frustration.

When you inevitably encounter a technical problem, you must walk a fine line and apply the "struggle first, but not forever" rule. Do not immediately ask for help. A huge part of your job as an engineer is to be a professional problem-solver. Before you ask for help, spend a reasonable amount of time trying to solve it yourself. Look for the documentation on the company's internal network. Read the relevant specifications. Use the help files in the software. Do not struggle in silence forever, however. The second biggest mistake a new hire can make is to waste an entire day stuck on a problem that a senior engineer could have solved in five minutes. This shows poor judgment.

The One-Hour Rule: A good rule of thumb is the "one-hour rule." Struggle with the problem on your own for a solid hour. If you are still completely stuck after that, it is time to ask for help.

The Amateur's Way: "I'm stuck. How do I do this?" (This signals helplessness and puts the entire burden on the mentor).

The Professional Approach: When you do ask, you must demonstrate that you respect the other person's time by proving you've already put in the effort. Do not simply declare defeat.

The Professional's Script:

Frame your question as a three-part brief. Walk up to your mentor and say:

(State Your Goal): "Hi Dave, quick question. I'm trying to apply the correct GD&T callout for this feature on the new bracket."

(State What You've Tried): "So far, I've read through the company drafting standard, Section 4.2, and I've tried using both the position and profile tolerance tools to control it."

(State Your Specific Question): "I'm still not sure which one is most appropriate for this specific application to control the interface. Could you point me in the right direction?"

(Why It Works): This script is a game-changer. It transforms you from a helpless new hire into a proactive junior colleague. It shows respect for your mentor's time, it demonstrates your diligence, and it allows them to provide a quick, targeted answer instead of a long, foundational lecture. It is the fastest way to earn the respect of the senior engineers on your team.

Under-Promise and Over-Deliver: This is a simple rule that will build immense trust with your manager. When you are asked for a time estimate, be realistic and even a little conservative. It is always better to say "I should have that done by Thursday" and deliver it on Wednesday, than to promise it by Tuesday and be late. A reputation for reliability is the most valuable currency you can earn.

The Manager Relationship: The Art of "Managing Up"

Your relationship with your direct manager is the single most important factor in your early career success. Your goal is to establish a professional, proactive relationship built on clear communication.

Schedule a Formal Kick-off Meeting: Your manager is busy. Take the initiative to schedule a 30-minute meeting within your first week or two to discuss your role and their expectations.

<u>Ask the Three Magic Questions:</u>

"What does a successful first 90 days look like for someone in my position?"

"What is the best way for me to communicate with you? Do you prefer email, instant messenger, or a quick drop-by?"

"What are the most important documents I should be reading to get up to speed?"

Before the meeting ends, ask to set up a recurring 15- or 30-minute one-on-one meeting, perhaps every other week. This is incredibly important. It gives you a dedicated, scheduled time to ask questions and provide status updates, so you don't have to constantly interrupt them.

By the end of your first month, you will have gotten your first "win." You will have proven that you can be trusted with a task, and you will have established a professional, proactive relationship with your manager. You are no longer just a new hire; you are a contributing member of the team.

Finding Your Place (Days 30-90)

By now, you have a handle on your core responsibilities. You know your teammates, you understand the rhythm of the work, and you have successfully delivered on your first few tasks. This is where you transition from being a reactive new hire who is simply executing tasks, to being a proactive engineer who is thinking about the bigger picture. Your goal in this phase is to expand your influence, to build your network, and to make yourself an indispensable part of the team.

Your Action Plan for Making an Impact

Take Deep, Personal Ownership of Your Work: Treat every task, no matter how small, as if you are the sole person responsible for its success. Think ahead. Anticipate problems. Don't wait to be told what to do next. If you finish your primary task on Wednesday, don't just wait for your next one-on-one on Friday. Go to your lead and ask, "I've finished the analysis on the bracket. While I'm waiting for the peer review, is there anything else I can help with?"

Build Your Network with Purpose: Your goal is to have numerous people outside of your immediate team who can vouch for you and your impact. This is not about office politics; it is about professional relationship-building. Set up 15-minute "informational interview" meetings with senior engineers or managers of teams you find interesting.

This can be done by sending a simple, respectful email: "Hi [Name], My name is Alex, I'm a new engineer on the structures team. I'm really interested in the work your team does in thermal analysis. I was wondering if you might have 15 minutes to spare in the coming weeks to tell me a little bit about your work. I'm just trying to learn as much as I can about the program."

Don't just talk about yourself. Ask them about their journey: "How did you get started in this group? What have been some of your favorite projects? What advice do you have for someone just starting out in this industry?" People love to share their stories, and it builds a genuine connection. This makes you a known, respected name across the organization.

Become More "T-Shaped": Ask your lead engineer if you can sit in on a meeting for a different team, like the software team or the test team, just to listen and learn. Understanding the challenges of the software team will make you a better mechanical engineer, and vice versa. This shows a level of curiosity and systems-level thinking that is rare and highly valued.

The Foundation is Laid

The first 90 days of your career are a unique and fleeting period. It is a time of immense learning, of frequent confusion, and of profound professional growth. The journey from the nervous excitement of your first day to the quiet confidence of being a trusted, contributing member of the team is a significant one. By the end of this initial phase, the feelings of being overwhelmed and useless should be replaced by a sense of belonging and purpose. You have learned the language, you have mapped the terrain, and you have gotten your first win.

It is crucial to recognize what you have truly accomplished. You have not just learned a new piece of software or completed a few tasks. You have begun to build the single most valuable asset of your career: your professional reputation. Through the small, consistent acts of your daily work, like showing up prepared, asking smart questions, taking ownership, and flawlessly executing your first assignments, you have sent a powerful message to your colleagues and your leaders. You have shown them that you are reliable, that you are coachable, and that you are a professional who is serious about their craft.

You have laid the foundation. The work you did in these three months, establishing a strong relationship with your manager, proving your diligence, and building your initial network, is the bedrock upon which the rest of your career will be built. You have earned a measure of trust, and that trust is the currency that will grant you access to more interesting, more challenging, and more rewarding work in the months and years to come.

The journey is far from over. In the next chapter, we will move beyond the initial onboarding and delve into the deeper, more subtle "unwritten rules" of the corporate culture. However, the hardest part is over; you have successfully navigated the transition from the academic world to the professional one. You have proven you belong. Remember that as a new engineer, your most important ability is your reliability. Be the person who does what they say they will do. Master this, and everything else will follow.

Chapter 25

The Unwritten Rules of a Defense Career

You've survived your first 90 days. You have a handle on your initial assignments, you know your teammates' names, and you've established a rhythm with your manager. You have successfully passed the initial test; the company has confirmed that you have the technical horsepower they hired you for. Now, a longer, more subtle, and arguably more important evaluation begins. This is the phase where you move beyond simply demonstrating your technical skills and start demonstrating your professional wisdom. It is the process of integrating into the unique culture of a large defense contractor.

Your engineering degree taught you the laws of physics, but it did not teach you the laws of large, high-consequence organizations. Your technical skills got you the job, but your ability to understand and navigate the corporate culture with the unwritten rules of the road will determine your long-term success and the trajectory of your career. It is the difference between being a competent engineer who completes their tasks and being a respected, influential engineer who is sought out for the most challenging and rewarding projects.

Every company has its own culture, but the established world of defense engineering has a set of powerful "unwritten rules" that are remarkably consistent across the industry. These rules are rarely written down in an employee handbook or discussed in an onboarding brief. They are the norms, the expectations, and the communication styles that have evolved over decades of high-stakes, high-consequence work. They are the professional DNA of the industry.

Understanding these rules is the key to moving from a competent new hire to a respected and effective member of the team. Ignoring them, even with the best of intentions, can lead to frustration and stall your career growth, even if you are a brilliant technical engineer. This chapter is your guide to these five critical unwritten rules, the insider's playbook for navigating the human and organizational side of defense engineering. Think of this as learning the physics of the organization.

Rule #1: Process is King

In the commercial tech world, the celebrated mantra is "move fast and break things." It is a philosophy born of the digital realm, a world where the consequences of failure are low and the speed of iteration is necessary for beating competition. If a new social media app has a bug, you can push a software patch the next day. The user is momentarily inconvenienced, the stock may dip for a day, but the world moves on. This is a valid and powerful model for that environment. It is, however, a mindset that you must shed at the door of a defense contractor.

In the world of aerospace and defense, the guiding principle is precisely the opposite: "Move deliberately and ensure nothing ever breaks." This is not because the industry is inherently slow or resistant to change; it is because the stakes are unimaginably high. The systems you build must work flawlessly, the first time and every time, for decades, often in the most extreme environments imaginable. You cannot issue a "software patch" for a missile that is in mid-flight. You cannot recall a submarine with a structural flaw from the bottom of the ocean. You cannot ask a pilot to "turn it off and on again" when their flight control system fails at Mach 2. The consequences of failure are measured not in lost revenue, but in mission failure and, potentially, in human lives.

This stark reality has led to the development of a highly structured, rigorous, and disciplined engineering process. This is the world of formal, milestone-based design reviews (PDR, CDR), where you must stand before a room of senior experts and customers and prove, with data, that your design is sound. It is the world of meticulous Configuration Management

(CM), a system where every single change to a design, no matter how small, is formally documented, reviewed, and approved before it can be implemented. It is a world of deep, foundational emphasis on Verification and Validation (V&V), where every single requirement is tracked and proven, through analysis and testing, to have been met.

For a new engineer, especially one accustomed to the fast-and-loose style of a university project where you might pull an all-nighter and make a major design change the day before the presentation, this can feel like bureaucracy. It can feel like a series of frustrating roadblocks designed to slow you down. It is not. See this as the accumulated wisdom of decades of high-consequence engineering. Generations of engineers designed this system after seeing the catastrophic results of a missed requirement or an undocumented change. It is the very DNA of a culture of safety and reliability.

Your first and most important challenge as a new engineer is to learn, respect, and ultimately embrace this process. Do not try to be a maverick who circumvents the system to save time. You will be seen not as innovative, but as reckless and unprofessional. The engineer who skips a step in the drawing release process because they are in a hurry is not a hero; they are a liability. Your value is not in creating your own clever workflow; it is in executing the proven, established workflow with excellence and integrity. The true creativity in this industry comes from solving an impossible technical problem *within* the disciplined framework of the process, not from fighting against it.

How to Master the Process

Understanding the importance of process is one thing; learning to navigate it is another. As a new engineer, you can take several concrete steps to quickly master your company's way of doing business and build a reputation for diligence.

Find the "Process Gurus": On every team, there are one or two mid-career or senior engineers who are the unofficial experts on the company's internal processes. They know the correct way to fill out an engineering change request, they know the unwritten rules of the drawing release system, and they know who to call in manufacturing to get a question answered. Identify these people early. When you are given your first formal task, politely ask for their guidance: "Hi Dave, I'm working on my first drawing release. I've read the company standard, but would you mind taking a quick look at my paperwork before I submit it to make sure I haven't missed anything?" This single act shows humility, respect for the process, and a commitment to getting it right the first time.

Treat Documentation as Engineering: This is a crucial mindset shift. Do not treat the paperwork and documentation as an annoying administrative task you do *after* the "real" engineering is finished. The documentation *is a part of the engineering*. It is a permanent record of your work, but also your process of finding the chosen solution. Write your analysis reports with the same rigor and attention to detail that you used for the analysis itself. A well-written, clear, and concise report that a future engineer can pick up and understand five years from now is a mark of a true professional and a profound service to the program.

Leverage your Peers: A common mistake new engineers make is to complete a formal process in isolation, submit it, and then have it rejected for a reason they could have known. The professional's secret is to "pre-coordinate." Before you formally submit a major change request, walk the draft over to the key stakeholders (like the lead from the manufacturing or stress team) and get their informal buy-in first. A 5-minute conversation beforehand can save you two weeks of rework on a rejected form. It shows respect for other disciplines and a strategic understanding of how the organization actually works.

Rule #2: Your Reputation is Your Most Valuable Asset

Because programs last for years and careers can last for decades at a single company, your reputation is everything. The defense engineering community within a large company, and even across the entire industry, is surprisingly small and interconnected. The lead engineer on your next project five years from now might be someone who only knows you from a brief interaction you had as a new hire. The opinion they form of you today, of your reliability, your professionalism, and your character, will follow you for a long time. It is the invisible currency of your career.

Your reputation is a combination of your technical competence and your professional character, and it is built not in grand, heroic gestures, but in the small, consistent actions of your daily work. It is the sum of every deadline you meet, every meeting you come to prepared for, every technical question you answer honestly, and every interaction you have with your colleagues. A brilliant engineer who is unreliable or difficult to work with will quickly find themselves sidelined, relegated to purely technical tasks with little influence. In contrast, a competent, trustworthy, and collaborative engineer will be given ever-increasing levels of responsibility and will be sought out for the most interesting and challenging projects.

How to Build Your Reputation

A great reputation is not built by accident. It is the result of deliberate, consistent professional habits. Here are the day-to-day actions that will build your reputation as a trusted and reliable engineer.

The most important metric for reliability is your *"say-do" ratio*. Strive for it to be 1.000. This means that every single time you say you will do something, you do it. Do not make casual promises. If you tell a colleague, "I'll send you that data by the end of the day," put it in your calendar and make sure it gets done. This small, consistent act of keeping your word will build a foundation of trust faster than anything else.

Provide early warnings. You will inevitably run into problems that will cause you to miss a deadline. A reliable engineer does not wait until the deadline to announce a problem. As soon as you identify a significant risk to your schedule, you must provide an "early warning" to your lead engineer. Frame it professionally: "Hi Sarah, I wanted to give you a heads-up on my analysis task. I've run into an unexpected issue with the model convergence, and I think it might put the Thursday deadline at risk. I'm trying a new approach now, but I wanted you to be aware of the situation." This is not a sign of failure; it is a sign of professional maturity.

Walk the floor. Do not be an engineer who lives only in their cubicle. Once a week, take a walk down to the manufacturing floor or the integration lab. Talk to the technicians who are building and testing the hardware you designed. Ask them questions. "Is this feature difficult to machine? Is there a better way I could have designed this for assembly?" This shows respect for their incredible expertise and will earn you a reputation as a collaborative, team-oriented engineer. Don't just talk to the techs, though; get hands-on. Learn their process to gain a deeper understanding of why they perform tasks in a certain way or why something is challenging. Working on the floor will make you a much better designer with the context and perspectives that you gain.

Own your mistakes, instantly and completely. You will make a mistake. You will run an analysis with the wrong boundary condition. You will order the wrong part. The mistake itself is not what defines your reputation; the way you react to it is. The amateur's instinct is to hide it or deflect blame. The professional's response is to take immediate, complete ownership. The moment you realize an error, you must go directly to your lead engineer and say, "Sarah, I've made a mistake. I just discovered I used the wrong material property in the analysis I sent you yesterday. It's my fault, and I'm already rerunning it now. I will have the corrected results for you this afternoon." This is a terrifying conversation to have, but it is a career-defining one. An engineer who owns their mistakes is an engineer who can be trusted. An engineer who hides them is a liability who cannot.

Rule #3: The Meeting is the Work

A common complaint from new engineers, especially those accustomed to the long, focused coding sessions of a university computer science program, is the sheer number of meetings. It's easy to feel like they are a distraction from your "real" engineering work that happens at your desk. This is a fundamental misunderstanding of large-scale systems engineering. On a large, complex program, the meetings *are* a critical part of the work.

Your job is not just to design your component in isolation; it's to ensure your component integrates perfectly with dozens of others, each designed by a different team. The meeting, whether it's a formal design review, an interface control meeting, or a simple team stand-up, is the primary mechanism where that integration happens. It is where requirements are negotiated, where interfaces are defined, where risks are identified, and where collective, cross-disciplinary problems are solved. It is the circulatory system of the program.

You must master the art of the meeting. Come to every meeting prepared. Understand the agenda. Have your data and your analysis ready. When you speak, be concise, clear, and data driven. State your point, provide your evidence, and then stop talking. Rambling is a sign of a disorganized mind. When you are not speaking, listen actively. Try to understand the challenges and constraints that other teams are facing. Proving you can contribute effectively in a collaborative meeting is just as important as the brilliant analysis you do at your desk.

How to Master the Meeting

You can dramatically increase your value to the team by being an exceptional meeting participant, not just a passive attendee.

The "5-Minute Prep": Before any meeting, take just five minutes to prepare. Read the agenda. Open the documents or the presentation that will be discussed. Think of one or two intelligent questions you need to ask; don't ask questions for the sake of being heard. They should be relevant and important. Compile your updates since the last meeting, including but not limited to what is finished, what is late, and what are unforeseen challenges that you are facing. This simple act will put you in the top 10% of meeting participants and ensure you can contribute effectively.

The "Offline" Follow-Up: A common mistake is to try to solve a deep technical problem in a meeting with ten people. This is an inefficient use of everyone's time. A better approach is to identify the problem in the meeting and then take the detailed conversation "offline." You can say, "That's a really important technical point. Rather than derail this meeting, perhaps Dave and I can grab 15 minutes at the whiteboard after this to dig into the details and report back to the team." This shows respect for your colleagues' time and a focus on efficient problem-solving.

Be the Scribe: A simple but incredibly valuable role in any meeting is to be the scribe. Take clear, concise notes, and be sure to capture the action items: who is responsible for what, and by when. Sending out a brief summary email after the meeting with the key decisions and action items is a powerful way to demonstrate your value and your emerging leadership and organizational skills.

Rule #4: Master "Matrixed" Communication

In most large defense companies, you are part of a "matrixed" organization. While confusing at first, the matrix provides a powerful and efficient way to manage large pools of talent. It means you essentially have two bosses, each with a different role in your career, and you must learn to communicate effectively with both.

Your functional manager (e.g., the Head of the Mechanical Engineering Stress Analysis Department) is your "home" manager. They are your long-term career advocate. They are responsible for your training, your skill development, your performance reviews, and for assigning you to different programs. They are focused on your growth as a mechanical engineer.

Your program manager or lead engineer on the specific program you are assigned to is your day-to-day boss. They are responsible for your tasks, your schedule, and your technical execution on that specific project. Their focus is on the success of the program.

Sometimes, the goals of these two leaders can be in a healthy tension. The program lead wants you to finish your task as quickly as possible. Your functional manager wants you to do it in a way that develops a new skill. You must learn to keep both of these stakeholders informed. Your program lead needs to know the daily status of your technical work. Your functional manager needs to know about your career goals, your desire for training, and any challenges you are facing. Never try to play one off the other. Open and honest communication with both sides of the matrix is essential for navigating the organization and ensuring that both your project work and your long-term career goals are being met.

How to Navigate the Matrix

Successfully navigating the matrix requires a deliberate and proactive communication strategy.

Establish a Rhythm with Both Managers: As we discussed in the "First 90 Days" chapter, you should have a recurring one-on-one with your program lead to discuss your day-to-day tasks. You should *also* have a recurring (perhaps monthly) one-on-one with your functional "home" manager. Use this meeting specifically to discuss your long-term career goals, your training needs, and the skills you want to develop.

Never Blindside Your Functional Manager: Your program lead will give you your annual performance review input, but it is often your functional manager who formally delivers it and makes the final decision on your rating and salary. Therefore, it is critical that you keep your functional manager aware of your accomplishments. Once a month, send them a brief, bulleted email that says, "Hi [Manager's Name], just wanted to keep you in the loop on my progress on the XYZ program this month. Key accomplishments included..." This ensures that when it's time for your performance review, they have a complete picture of your contributions.

When in Doubt, Over-Communicate: If you are unsure if a piece of information is relevant to your program lead or your functional manager, err on the side of over-communicating. A simple "FYI" email that keeps them in the loop is always better than them being surprised by a problem or a success that they should have known about.

Rule #5: Sustainability is a Discipline

The mission is urgent, the stakes are high, and the problems are fascinating. This combination creates a dangerous trap: the belief that you must work 60 hours a week to be a "good" defense engineer. While there are times for "surges"—proposal deadlines, flight tests, or critical milestones—a career in this industry is a marathon, not a sprint.

Burnout is a risk multiplier. A tired engineer makes mistakes. A tired engineer misses a critical tolerance in an analysis or clicks the wrong link in an email. The culture of defense takes this seriously. This is why many companies operate on a "9/80 schedule" (working 80 hours over 9 days to get every other Friday off). That off-Friday is not just a perk; it is a structural tool for mental recovery.

You must learn to disconnect. Be fully present when you are at work, but when you badge out, be fully present in your life. Building a sustainable rhythm, where you protect your sleep and your sanity, is the only way to maintain the high level of cognitive performance this job demands for the long haul.

Rule #6: Become a "T-Shaped" Engineer

The company hired you for your deep technical knowledge in your discipline, representing the vertical bar of the "T". You are a mechanical engineer, or a software engineer, or an electrical engineer. Expertise forms the foundation of your value, and your early career will be focused on making this bar as deep and as strong as possible.

But your long-term value, and your ability to progress into leadership roles (whether technical or managerial), lies in your ability to also develop the horizontal bar of the "T": a broad understanding of the adjacent disciplines. You must learn to speak the language of the other engineering disciplines. You must develop an appreciation for their challenges, their constraints, and their culture.

Be curious. Take the time to understand what the electrical engineers, software engineers, and test engineers on your team are working on. Ask them questions. When an EE talks about "signal integrity," ask them what that means. When a software engineer talks about "real-time operating systems," ask them what makes them different. Go to the manufacturing floor and talk to the machinists. Go to the integration lab and talk to the test technicians.

The mechanical engineer who understands the basics of circuit board layout and the constraints it places on their housing design will always be more valuable than the one who throws their design "over the wall." The software engineer who understands the physics of the system they are controlling will write better, safer, and more effective code. The engineers who can operate at the seams between the disciplines are the ones who solve the hardest and most important integration problems, and they are the ones who become the future Chief Engineers and Program Managers.

How to Build Your "T"

Becoming "T-shaped" requires a deliberate investment of your time and intellectual curiosity.

The "Lunch and Learn": Many companies have informal "lunch and learn" sessions where an engineer from another team presents their work. Go to them. Even if the topic seems unrelated to your work, it is a fantastic, low-effort way to absorb the language and the challenges of other disciplines.

Read the ICDs: The Interface Control Document (ICD) is the formal document that defines how your component connects to another component. When you are assigned a task, don't just focus on your part. Ask for the ICD and read it. Take the time to understand the requirements and constraints of the system that your part is plugging into. This is the first step to thinking at a systems level.

Find a Mentor in Another Discipline: Find a mid-career engineer in a different discipline (e.g., a software engineer if you are a mechanical engineer) and ask them to be an informal mentor. Take them to coffee once a quarter and ask them questions about their world. "What is the hardest part of your job? What do you wish the mechanical engineers on your team understood better?" This cross-disciplinary mentorship is an incredibly powerful way to build the horizontal bar of your "T."

From Individual Contributor to Influential Teammate

The five rules we have discussed, revering the process, cultivating your reputation, understanding the role of meetings, mastering the matrix, maintaining sustainability, and becoming T-shaped, are not just a list of tips. They are a holistic guide to professional maturity in the defense industry. They represent the transition from thinking like a student, who is focused on their individual performance, to thinking like a professional systems engineer, who is focused on the success of the larger enterprise.

Mastering these rules is how you build the invisible currency of a great career: trust. When you follow the process, you prove you can be trusted with high-consequence tasks. When you build a reputation for reliability, you prove you can be trusted with responsibility. When you master the meeting, you prove you can be trusted in a collaborative environment. When you communicate effectively within the matrix, you prove you can be trusted with complex organizational relationships. And when you become T-shaped, you prove you can be trusted to understand the big picture.

This foundation of trust is the prerequisite for everything that comes next. It is what will earn you the respect of your peers, the confidence of your leaders, and the opportunity to take on more challenging and more rewarding work. It is the key that unlocks the door to the career paths we will discuss in the next chapter.

As you move forward, you will have to make a conscious decision about the kind of engineer you want to be. Do you want to be the deep technical expert who solves the impossible problems? The inspiring leader who builds and mentors great teams? Or the technical manager who guides a program to success? Your ability to ascend any of these paths is not just a function of your technical skill, but of your mastery of these unwritten rules. They are the foundation upon which a remarkable career is built.

Chapter 26

Career Progression

You've successfully navigated your first year. You're a contributing member of your team, you understand the company culture, and you've started to build a reputation for being a reliable engineer. For the first three to five years, the path forward is the same for every engineer: build the trunk of your career tree. Your objective during this period is to build a deep and solid foundation of technical excellence. Become an expert in your discipline. Master the tools. Prove that you can deliver.

But a moment will come, typically around the five-year mark, when you arrive at your first major fork in the road. Having established your technical credibility, your career will begin to branch. This is not a simple choice between being "technical" or being a "manager." In a large defense contractor, your progression is a strategic choice between three distinct, equally valuable, and highly rewarding paths of leadership and influence.

Understanding the differences between these paths *early* is an essential act of strategic foresight. It allows you to consciously develop the skills, seek the experiences, and build the relationships that will propel you toward the future you want, preventing the common mistake of drifting into a role that doesn't fit your passions. This chapter is your long-range career map.

Table 1 A Summary of the Three Core Engineering Career Paths

Technical Fellowship Path *The Solver of Impossible Problems*	Program Leadership Path *The Conductor of the Orchestra*	Functional Management Path *The Cultivator of Talent*
Focus: Technology, Analysis, & Innovation	**Focus:** Technical Execution of a Program	**Focus:** People, Skills, & Resources
Manages: The "Why" & "How Good"	**Manages:** The "What" & "When"	**Manages:** The "Who" & "How"
Success Metric: The technical performance, integrity, and innovative edge of the system being designed.	**Success Metric:** The successful execution and delivery of a working system to the customer.	**Success Metric:** The long-term health and technical capability of their engineering department.
Typical Titles: Principal Engineer, Senior Principal Engineer, Technical Fellow, SME	**Typical Titles:** Lead Engineer, IPT Lead, Chief Engineer	**Typical Titles:** Section Manager, Department Manager, Director of Engineering

The Matrix and the Dual-Ladder System

To understand these three paths, you must first understand the fundamental organizational philosophy of a large engineering company: the "matrix organization." We've touched on this before, but its impact on your career is profound.

On one axis of the matrix are the Programs: the specific projects, like the F-35 or the new satellite, that are funded by the customer to deliver a product. These are focused on schedule and budget. On the other axis are the Functional Departments (sometimes called the "home office" or "Directorate"): the pools of expertise, like the "Mechanical Engineering Department" or the "Software Engineering Department." These are focused on technical excellence and talent development.

This structure created a classic problem: How do you promote and reward your most brilliant technical expert? In a traditional hierarchy, the only path upward is to become a manager of people, often taking your best problem-solver away from the very technical work they are best at. The solution was the "dual-career ladder," the system that gives rise to these distinct paths. It was designed to ensure that the company could retain and advance its most critical technical talent (The Technical Fellow) on a path parallel to its leaders of people (The Functional Manager). The Program Leadership path emerged as the nexus between the two, leading the technical work on the programs themselves. Understanding this structure is the key to seeing your career not as a ladder to be climbed, but as a strategic map to be navigated.

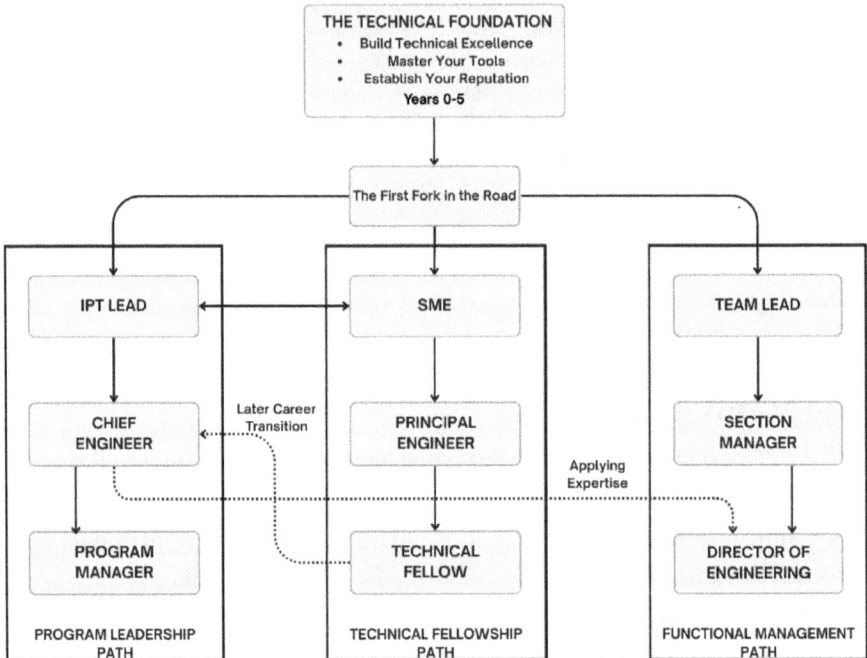

Figure 6 A visual map of the three primary career paths for a defense engineer, branching from a common foundation of early-career technical excellence.

Path 1: The Program Leader

The Program Leader operates in the dynamic, high-visibility gray area where most of the day-to-day engineering leadership happens. A Lead Engineer or a Chief Engineer is a technical manager. They are still deeply involved in engineering, but their primary job is to lead a multi-disciplinary team to a successful outcome. They are the ultimate owner of the technical solution. They make the final technical decisions, they manage the technical risks, and they are accountable to the overall Program Manager for the successful execution of the engineering work, on schedule and on budget.

What This Path Entails

A Program Leader lives at the intersection of the technical, programmatic, and the human. Their day is a constant balancing act. They must be fluent enough in a dozen different engineering disciplines to understand the challenges of each, and they must be able to translate those technical challenges into the language of schedule and cost for the program manager. They are the ultimate "T-shaped" engineers, with a deep enough understanding of all the disciplines to ask the right questions and to make intelligent, data-driven trade-off decisions.

They are the ones who must decide: "The structures team says we can save 10 pounds if we use a more expensive material, but the budget is already over. The software team says they can add a new feature, but it will delay the next test by two weeks. What is the right decision for the program?" They are the ultimate technical problem-solvers, but their problems are not just about physics; they are about the complex interplay of technology, time, and money.

A Day in the Life of a Program Leader

A Chief Engineer's day is a high-speed symphony of technical problem-solving and programmatic management. Their morning might start with an urgent TIM (Technical Interchange Meeting) to resolve a critical integration issue between the new sensor from a subcontractor and the team's own software. They are not there to solve the problem themselves, but to facilitate the conversation, to ask the right questions, and to drive the two teams to a consensus that is in the best interest of the overall system.

Their midday might be spent presenting a high-level summary of the program's top five technical risks to a panel of senior executives and customer representatives. They must be able to explain a complex technical problem in simple, clear terms and to confidently present the team's plan for mitigating the risk.

The afternoon is a rapid-fire series of one-on-ones with their Integrated Product Team (IPT) leads, getting status updates, removing roadblocks, and making the dozens of small technical decisions needed to keep the program on schedule. They are the ultimate technical decision-maker, the person who must balance risk, cost, and performance to deliver a working system.

Is This Path For You?

This path is for the engineer who is a master of their own discipline, but who is even more fascinated by how all the disciplines fit together.

You have a strong technical foundation, but you are also skilled at seeing the "big picture." You are the person who is always asking "why" and who is interested in how your component affects the system as a whole.

You are a "T-shaped" engineer. You can communicate effectively with mechanical, electrical, and software engineers because you have taken the time to learn the basics of their world.

You enjoy the challenge of making a complex system work on schedule and on budget. You are a pragmatist who is comfortable with the idea that "perfect is the enemy of good enough."

You are comfortable making difficult technical trade-off decisions with incomplete information. You have the confidence to make a judgment call based on the best available data and to take ownership of that decision.

How to Prepare

The path to program leadership is a journey from being the owner of a component to being the owner of a system.

Become a Strong, Cross-Disciplinary Engineer: The program leader must speak multiple technical languages. As we discussed, you must build the horizontal bar of your "T." Spend time with the other disciplines on your team. Buy the software engineer coffee and ask them about their biggest challenges. Your goal is to understand their world.

Volunteer to be an IPT Lead: The classic entry point into this track is to become the Integrated Product Team (IPT) Lead for a small part of the project. This gives you "ownership" of a specific piece of the program (like the "wing structure IPT") and teaches you how to manage the technical work, schedule, and cost. It is your first taste of being a true technical manager.

Find a Mentor Who is a Chief Engineer: Seek out a successful Chief Engineer on a program you admire. Ask them for an informational interview. Learn how they think, how they make difficult trade-off decisions, and what they look for in a good lead engineer.

Master the Language of Risk and Systems: Program leadership is the art of managing technical risk. Actively participate in risk-management discussions on your team. Seek out training in Systems Engineering and Model-Based Systems Engineering (MBSE) to learn the formal language and processes for managing complex systems.

Path 2: The Functional Manager

Functional Management represents the most traditional "manager" role, often referred to as "the home office" or the "discipline chief." These leaders manage a department of engineers based on their discipline (e.g., the "Mechanical Engineering Stress Analysis Group," the "RF Engineering Department," or the "Software Engineering Directorate"). Their primary job is not to manage a single program, but to manage the company's most precious resource: its people. Their customer is the program, and their product is a team of highly skilled, well-trained, and motivated engineers that they can deploy to meet the program's needs.

What This Path Entails

A functional manager is a servant leader, a mentor, and a business strategist all rolled into one. Their world is not one of CAD models and circuit diagrams, but of one-on-one meetings, budget spreadsheets, and strategic human capital planning. They are the ones who are responsible for hiring new talent, for interviewing the students who have read this book, and for building a team with the right mix of skills to meet the company's future needs. They are the ones who ensure their engineers have the training and the tools like software licenses and lab equipment that they need to succeed.

They are also the ones responsible for conducting the annual performance reviews that determine raises and promotions. This is one of their most important and difficult jobs: to provide honest, constructive feedback, to celebrate successes, and to help an engineer who is struggling to get back on track. They are the career counselors, the mentors, and the advocates for the engineers in their group. When a new program needs a stress analyst, the program manager comes to the functional manager to request a person. The functional manager, knowing the skills, the performance history, and the career goals of everyone in their department, will select the right person for the job, a decision that can shape both the program's success and the engineer's career.

A Day in the Life of a Functional Manager

A functional manager's day is a tapestry of human interaction and strategic planning, often scheduled in 30-minute blocks. Their morning might start with a one-on-one meeting with a junior engineer who is struggling with a difficult technical task, where the manager's role is not to solve the problem, but to coach the engineer, to ask the right questions, to build their confidence, and to connect them with the right senior mentor who can help with the technical details.

Mid-morning might be a long, difficult budget meeting with other department heads, where they have to advocate for funding to buy new, expensive software licenses for their team or to send a high-potential engineer to a critical training conference. This is where they act as the business leader for their group, making a data-driven case for why this investment is critical for the company's long-term success.

The afternoon could be spent screening resumes and conducting phone interviews for a new open position in their department, a crucial task for the long-term health of the team. They are not just looking for technical skills; they are looking for the right attitude, the right cultural fit, and the potential for future growth.

The day might end with them reviewing the annual performance goals for all fifteen engineers in their group, ensuring each person has a clear and achievable path for growth in the coming year. Their work is a constant, demanding, and incredibly rewarding process of cultivating the human talent that is the true engine of the company.

Is This Path For You?

This path is for the engineer who discovers, often to their own surprise, that they derive the most satisfaction from helping others succeed. It is for the person who finds that they enjoy mentoring the intern on their team even more than they enjoy doing their own analysis.

You are passionate about mentoring and teaching. You are the person your teammates naturally come to for advice, and you take pride in seeing them learn and grow.

You are a natural organizer. In your team projects, you find yourself naturally taking on the role of the person who organizes the tasks, who ensures everyone is on the same page, and who leads the final presentation.

You are interested in the "business" of engineering. You are curious about how the department is run, how budgets are allocated, and how the company makes strategic decisions about hiring and training.

You are a natural communicator and an empathetic listener. You are comfortable having difficult conversations, you can mediate disputes, and you are genuinely invested in the personal and professional well-being of the people around you.

How to Prepare

The path to functional management is a gradual one, built on a foundation of technical excellence and a demonstrated aptitude for leadership.

Excel Technically First: You cannot lead engineers without first earning their respect as a highly competent engineer in your own right. For your first 3-5 years, your sole focus should be on technical excellence.

Seek Out Mentorship Opportunities: Volunteer to be the official mentor for a new intern or a new hire. This is a low-risk way to gain experience in coaching and developing talent, and to see if you genuinely enjoy it. Your manager will notice this initiative.

Express Your Interest: In a one-on-one with your own functional manager, have a direct conversation. Tell them that you are interested in exploring the leadership path *in the long term.* Ask them, "What are the key skills and experiences I need to develop over the next few years to be considered for a team lead or section manager role?" A good manager will be thrilled to help you build that roadmap.

Take the "Business of Engineering" Training: Most large companies have a suite of internal training courses on leadership, conflict resolution, project management, and finance. Sign up for them. These formal courses will give you the language and frameworks to complement your informal experience.

Path 3: The Technical Fellow

The Technical Fellowship offers a path of pure, deep technical expertise. The Technical Fellowship track is a parallel ladder designed to ensure that the company's most brilliant minds can continue to advance in seniority and compensation without ever having to become a manager. Technical Fellows are the company's legendary experts, the "graybeards" who are consulted on the most critical and seemingly impossible technical challenges. When a program is facing a catastrophic failure, when a new technology is not working as promised, or when the company is bidding on a new, high-risk program, the Chief Engineer calls the Technical Fellow.

What This Path Entails

A Technical Fellow is a deep, intellectual problem-solver. Their days are spent in the world of analysis, simulation, foundational research, and deep, focused thought. They are often given a level of freedom and autonomy that is rare in a large corporation, allowed to pursue the problems they find most interesting and most critical to the company's future. They are the ones who are called in to be the "murder board" in a critical design review, the ones who are expected to ask the impossibly hard questions and to find the one fatal flaw in a design that everyone else missed.

They are the ones who are leading the company's most advanced Internal Research and Development (IRAD) projects, inventing the next-generation technologies that will give the company a competitive advantage five or ten years in the future. They are the ones who are representing the company at major academic conferences, who are publishing papers, and who are serving as the bridge between the world of academia and the world of applied engineering. And, most importantly, they are the ultimate mentors for the young engineers, the keepers of the corporate technical memory, the ones who are tasked with passing on their decades of hard-won knowledge to the next generation.

A Day in the Life of a Technical Fellow

A Technical Fellow's day is a journey into the deepest and most difficult technical problems the company faces, often unconstrained by the daily rhythm of a single program. Their morning might be spent in a secure, windowless room, reviewing the highly classified test data from a recent hypersonic flight, trying to understand an anomalous temperature reading that no one else can explain. They are not just looking at the data; they are deconstructing the entire test, questioning the sensor placement, the data acquisition system, and the underlying physics of the problem.

Their afternoon might be spent at a whiteboard, mentoring a team of young engineers who are working on a new AI algorithm. The Fellow is not there to write the code; they are there to challenge the team's assumptions, to ask the probing questions about the training data, and to guide them to their own "aha" moment.

The end of their day might be spent quietly reading a new, obscure academic paper on a cutting-edge material, constantly sharpening their own sword and looking for the next breakthrough. They might be preparing a presentation for the company's Chief Technology Officer on a new area of research the company should invest in. They are the company's technical conscience, operating on a timescale of years and decades, not days and weeks.

Is This Path For You?

This path is for the engineer whose greatest passion is the technology itself. You love the "hard problems." Your greatest satisfaction comes from cracking a difficult analysis that no one else could, or from finally understanding a complex physical phenomenon.

You are a lifelong learner. You are the person who is always reading technical books and papers, who is always taking online courses to learn a new tool, and who is insatiably curious about why things work the way they do.

You would rather mentor than manage. You love teaching and sharing your knowledge, but you would rather spend your day mentoring a junior engineer on a complex technical problem than sitting in a budget meeting.

You have deep technical patience. You are comfortable with the idea of spending years, or even a decade, becoming a true, world-class expert in a single, narrow, and incredibly complex niche.

How to Prepare

The path to becoming a Technical Fellow is a long and demanding one. It is a marathon of deep, focused, and continuous learning.

Go Deep, Not Broad: In your early career, find a specific, high-demand technical niche on your program and strive to become the undisputed expert on your team. Become the "vibrations SME," the "Python scripting guru," or the "RF antenna modeling expert."

Pursue a Graduate Degree: Use your company's tuition assistance program to pursue a Master's degree or, ideally for this track, a Ph.D. in a specialized technical field. This is the formal credential that validates your deep expertise.

Publish and Present: An expert's knowledge is only valuable if it is shared. You must build your reputation by publishing your work, both internally in company-wide technical journals and externally in peer-reviewed papers and at major industry conferences (like the AIAA SciTech Forum).

Find a Fellow as a Mentor: Seek out one of the Technical Fellows in your company. Ask them for a 30-minute informational interview. Learn about their career path, the critical projects they worked on, and the advice they have for an aspiring Fellow. This mentorship is invaluable.

A Map, Not a Cage

We have now explored the three primary paths of progression for an engineer in the defense industry: the Functional Manager who cultivates talent, the Technical Fellow who solves the impossible problems, and the

Program Leader who conducts the technical orchestra. Having this map in mind from the beginning of your career is an incredible advantage. It gives you a framework for your ambitions, it helps you understand the roles of the leaders around you, and it provides a clear set of signposts to look for as you develop your own skills and passions.

It is crucial to understand, however, that this is a map, not a cage. These paths are not rigid, mutually exclusive silos, and your career is not a one-way street. The reality of a long and rewarding career is often far more fluid and interesting. Many of the most successful and respected senior leaders have spent time on two, or even all three, of these paths. A brilliant Technical Fellow might be asked to lead a critical program for a few years because of their unique expertise. A successful Chief Engineer might transition later in their career to a Functional Management role to focus on mentoring the next generation.

The most important takeaway is that having a plan is the best way to ensure growth, but it is equally important to remain open to changing interests and unexpected opportunities. You may start your career absolutely convinced that you want to be a Technical Fellow, only to discover five years in that you have a hidden talent and passion for leading teams. That is not a failure; that is a discovery. The foundation of deep technical excellence you built in pursuit of the first path is the exact same foundation that will make you a credible and effective leader on another.

The one constant across all three of these paths is the need for continuous learning. Whether you are striving to be the ultimate technical expert, the most effective cultivator of talent, or the most knowledgeable program leader, your growth will be fueled by your commitment to deepening and broadening your knowledge. This brings us to the final piece of your long-range career strategy: the powerful and non-negotiable role of continuing education.

Chapter 27

Continuing Education

You've landed the job, you're navigating the culture, and you have a clear map of the potential career paths ahead of you. You have successfully built the foundation of your engineering career. The final step in ensuring your long-term success, and in preparing yourself for the leadership paths we've discussed, is to commit to being a lifelong learner. In an industry defined by a relentless pace of technological change, where the adversary is constantly innovating, the engineer who stops learning the day they graduate is the engineer who will, with absolute certainty, become obsolete. The skills that got you your first job will not be the skills that get you your last promotion.

The most successful and respected engineers, whether they are Technical Fellows, Program Leaders, or Functional Managers, are the ones who view their education not as a finite period in their youth, but as a continuous, career-long pursuit. They are insatiably curious. They read voraciously. They actively seek out new skills and new challenges. They understand that their most valuable asset is not what they know today, but their ability to learn what they will need to know tomorrow. They are not just engineers; they are professional students of their craft.

One of the most powerful and structured ways to formalize this commitment is by pursuing a graduate degree. And the single greatest, yet often underutilized, benefit offered by large defense contractors is their Tuition Assistance Program. This is your chance to earn a highly valuable Master of Science degree, often for free, while you continue to work and gain invaluable professional experience. This chapter is your guide to strategically leveraging this benefit, as well as the other pillars of professional development, to accelerate your career.

The Master's Degree

The decision to pursue a master's degree is a significant one, requiring a substantial investment of your time and energy over several years. It is not a step to be taken lightly. It will involve late nights of studying after a long day of work and sacrificing weekends to write papers and prepare for exams. However, when approached with a clear strategy and a specific goal in mind, it can be the single most powerful accelerator for your career. It is the formal, structured path to moving beyond the broad, foundational knowledge of your undergraduate degree and into the deep, specialized expertise that is the hallmark of a senior engineer.

The Strategic "Why"

A master's degree is more than just another line on your resume; it is a strategic career accelerator. In an industry filled with high-achievers, where a bachelor's degree is the baseline, a master's is a powerful differentiator that serves several key purposes. It signals a level of dedication, intellectual curiosity, and theoretical understanding that sets you apart from your peers.

Deepening Your Technical "Spike": This is the primary and most important reason for engineers on the Technical Fellowship track. A master's degree allows you to specialize in a high-demand field (like control systems, RF engineering, or materials science) at a depth that is simply impossible in a broad undergraduate curriculum. It is where you move from being a user of a tool (like FEA) to an expert who understands the deep theory behind it. This is the official stamp that says you are a subject matter expert in training, and it is often a prerequisite for the most advanced and challenging technical roles in the company.

Unlocking Higher-Level Positions: For many senior technical and leadership roles, a master's degree is a preferred, and sometimes required, qualification. As you climb the career ladder, you will be competing against other highly talented and motivated engineers. A master's degree can be the deciding factor that gives you the edge.

Keeping You on the Cutting Edge: Your undergraduate engineering program taught you the fundamentals, the laws of physics that don't change. A good graduate program teaches you the state-of-the-art. It exposes you to the latest research, the most modern tools, and the newest techniques in your field, ensuring your skills remain relevant and valuable in a rapidly evolving technological landscape.

Increasing Your Earning Potential: Over the course of a career, engineers with a master's degree consistently earn more than those with only a bachelor's. This is a simple statistical fact. When the company is paying for the degree, the financial return on your investment of time and effort is effectively infinite.

How Leadership Views Your Graduate Study

A common fear for a new engineer is that pursuing a master's degree will be seen as a distraction, or that struggling to balance it with work will make them look like a poor performer. It is natural to worry: "Will my manager think I'm not focused on my job?"

In the established culture of the defense and aerospace industry, the opposite is true. Actively pursuing a technical graduate degree using the company's tuition assistance program is one of the most powerful, positive signals a young engineer can send to leadership.

Enrolling in a master's program tells your manager and the wider organization three critical things:

It signals ambition and initiative. You are not content with the baseline. You are actively investing your own time and energy outside of your 40-hour work week to improve your skills and become more valuable to the team.

It signals long-term commitment. Especially when combined with a service agreement, it shows the company that you see a long-term future with them. You are investing in skills that you intend to apply on their programs for years to come.

It signals that you are a high-potential employee. Managers are constantly on the lookout for their future leaders. An engineer who proactively seeks to deepen their expertise through graduate study is self-identifying as someone who is serious about their career and has the drive to succeed. It is often a key differentiator when it comes time for promotions and assignments to high-visibility projects.

Therefore, do not treat your graduate study as a personal chore you must hide. Treat it as a formal, positive part of your professional development. Keep your manager informed of your progress in your one-on-ones. This is not complaining about being busy; it is reporting status on a long-term strategic project that benefits both you and the company.

The Two Paths of Company-Funded Education

When it comes to continuing your education, large defense contractors typically offer two distinct and powerful pathways. The most common is the Part-Time Tuition Assistance Program, which is available to most engineers and represents a steady, long-term investment in yourself. The second, more competitive and immersive option, is the Full-Time Graduate Study Fellowship, which represents a massive, focused investment by the company in their highest-potential employees. Understanding the difference is key to choosing the path that is right for you.

Path 1: Part-Time Tuition Assistance

Nearly every major defense contractor offers a generous tuition reimbursement program. This is the workhorse of continuing education and the path that most engineers will take. While the specifics vary from company to company, they generally follow this model:

How it works: The company will pay for your tuition, fees, and sometimes even books, up to a generous annual cap (often in the range of 10,000–25,000 per year). This is often enough to cover the full cost of a part-time master's degree from a top university. During the program, you must maintain a certain GPA in your graduate courses (usually a B average or higher) to be reimbursed.

In exchange for this massive investment in you, the company will typically require you to sign a service agreement. This means you must agree to remain an employee of the company for a certain period of time (usually one to two years) after your last reimbursed course is complete. If you choose to leave the company before this period is over, you may have to pay back a portion of the tuition assistance. This is a very fair arrangement, often referred to as "golden handcuffs."

The "How" for Part-Time Study:

Juggling a demanding full-time job and a graduate-level course is a marathon, not a sprint. Success requires not just intellectual horsepower, but discipline, excellent time management, and a clear strategy.

Do not start a master's program in your first six months, or even your first year, on the job. This is the single biggest mistake young engineers make. Your first year is for acclimating: getting your feet under you, mastering your core role, understanding the company culture, and establishing a sustainable work-life balance. Starting a graduate program before you have this foundation is a recipe for burnout and mediocre performance in both your job and your studies.

Your choice of a master's degree should not be a random continuation of your undergraduate major. It should be a deliberate, strategic decision made in direct consultation with your functional manager and informed by the career path you want to pursue.

For the future Technical Fellow: Pursue a deep, specialized technical degree. A Master of Science in Mechanical Engineering with a thesis in Solid Mechanics or a Master of Science in Electrical Engineering with a focus on Electromagnetics or DSP are excellent choices that build your "spike."

For the future Program Leader: Consider a broader, more interdisciplinary degree. A Master of Science in Systems Engineering is the perfect choice for this track. An MBA is generally not recommended for early-career engineers and is often seen as moving you "away from the work."

The temptation is to jump in and take a full course load to finish quickly. Do not do this. Start with a single course in your first semester. This will allow you to calibrate the workload and understand the real impact on your time. It is far better to succeed brilliantly in one course than to perform poorly in two. After a successful first semester, you can make an informed decision about whether you can handle a heavier load.

While a degree from a top-tier, nationally known university (like Stanford or Northwestern) carries significant weight, do not underestimate the value of a strong, local state university. Online programs have made top-tier schools more accessible, but a local option may provide a better rhythm and direct connections to other local engineers in your industry. This is another excellent topic for discussion with your manager and mentors. The company's tuition assistance program will give you the freedom to choose what's best for you.

Path 2: Full-Time Graduate Study Fellowships

This is an incredible opportunity for a company's highest potential early-career engineers. Many of the top defense contractors and national laboratories have highly competitive internal fellowship or "work-study" programs. They are designed to identify and fast-track their best and brightest, recognizing that a deep, immersive educational experience can forge a future leader. This is a massive, focused investment *by the company in you.*

How It Works: After working for the company for a year or two and establishing yourself as a top performer, you can apply for one of these prestigious internal programs. If selected, the company will essentially pay you to go back to school full-time for your master's degree. They will not only pay your full tuition and fees at a top university of your choice, but they will also pay you a reduced, but still substantial, full-time salary and maintain your benefits and security clearance. As with the part-time tuition assistance program, you sign a service agreement to return to the company for a set number of years after you graduate. Oftentimes, the service agreement is longer, but it will typically come with a significant promotion and raise waiting for you.

This is an incredible, life-changing opportunity. It allows you to be completely immersed in your studies, to participate in on-campus research with world-class professors, and to finish your degree in half the time it would take part-time, all while incurring zero debt. More importantly, being selected for a fellowship is a massive signal to the company's senior leadership. It publicly identifies you as a high-potential employee, placing you on the fast track for future technical and leadership roles.

The "How" for Full-Time Study

These programs are not advertised on the main career page. They are highly competitive internal opportunities, and you must build a strategic campaign to win one from your first day on the job.

Excellence in your first two years is non-negotiable given the inherently competitive nature of these programs. You must aim to build a reputation as a top 5% performer in your peer group. Your performance reviews must be stellar. You must be the junior engineer who is trusted with the hard problems and who consistently over-delivers.

From your first one-on-one, make your long-term educational goals known to your functional manager. Frame it as a desire to bring new, high-value skills back to the team. Ask them directly, "I am very interested in a deep specialization in control theory. Does our company offer any full-time fellowships or work-study programs for a master's degree that I could work toward?"

The best way to learn how to win is to talk to someone who has already won. Find a mid-career or senior engineer in your department who went through the fellowship program. Take them to coffee. Ask them about their experience, the application process, and what they believe made them a successful candidate.

Your functional manager's recommendation is the most important part of your application package. Your work in your first two years, your proactive communication about your goals, and your stellar reputation are all a means to an end: ensuring that when the application period opens, your manager will be your biggest advocate and write you a powerful, enthusiastic letter of support.

Part 2: Professional and Academic Certifications

While a master's degree is a powerful tool for deepening your theoretical knowledge, there are other, more focused ways to build and prove your expertise. Professional and academic certificates are often faster and more targeted than a full degree program, and they provide a clear, industry-recognized credential that you can add to your resume and your LinkedIn profile. A well-chosen certificate can be a powerful signal that you are a dedicated professional who is committed to mastering your craft. They are the tangible evidence of your practical skills and specialized knowledge.

Professional Certifications

In a world of digital engineering, your proficiency with your primary software tools is a direct measure of your productivity and your value to a team. A professional certification from a software vendor is proof that you are not just a casual user, but fluent and have invested the time to achieve a high level of mastery.

For Mechanical Engineers (e.g., CSWP): The most respected certifications for mechanical engineers are those related to CAD software. The CSWP (Certified SOLIDWORKS Professional) or its more advanced sibling, the CSWE (Certified SOLIDWORKS Expert), are prime examples. These are not simple multiple-choice tests. They are timed, practical exams where you are given a set of complex parts and drawings and are required to quickly and accurately model them, create assemblies, and troubleshoot design problems live in the software. Passing an exam like the CSWP is a definitive statement of your CAD skills. It proves to a hiring manager or your functional manager that you are fluent in the language of 3D modeling, that you are efficient, and that you understand the advanced features of the tool.

For Systems Engineers (ASEP and CSEP): The standard for anyone on the Program Leadership or Systems Engineering path is the Associate Systems Engineering Professional (ASEP) and the Certified Systems Engineering Professional (CSEP) from the International Council on Systems Engineering (INCOSE). The ASEP is meant for those just starting in a Systems Engineering role and provides credibility that you have built a foundation to work off of. The CSEP, however, is not a simple certification. It is a rigorous process that typically requires a minimum of five years of documented systems engineering experience, recommendations from other senior engineers, and passing a comprehensive, two-hour exam based on the official INCOSE Systems Engineering Handbook. Earning your CSEP is a major career milestone, a formal, industry-wide recognition of your expertise in the entire systems engineering lifecycle.

For Aspiring Program Leaders (PMP): The Project Management Professional (PMP) certification from the Project Management Institute (PMI) is the most recognized project management credential in the world. While the CSEP focuses on the *technical* management of a system, the PMP focuses on the *programmatic* management of a project: scope, schedule, and budget. For an engineer on the Program Leadership path, or one who wishes to eventually become a Program Manager, the PMP is an incredibly valuable asset that proves you speak the language of business and program management.

Academic Graduate Certificates

This is the powerful middle ground between a single training course and a full, two-year master's degree. A Graduate Certificate is a formal academic credential offered by a university, typically consisting of three to five graduate-level courses in a highly specialized technical area. For the engineer who wants to gain deep, focused knowledge in a specific niche without committing to a full master's program, they are a perfect solution.

These programs are a fantastic way to build a new "spike" of expertise. Because they are a bundle of actual graduate-level courses, the credits earned can often be applied toward a full master's degree later if you choose to continue. Many companies' tuition assistance programs will fully cover the cost of these certificate programs.

How to Choose a Track: The key is to choose a certificate that is a direct force multiplier for your current role or your desired future role. This is a strategic decision that should be made in consultation with your manager and mentors. Look for a program that aligns with the "megatrends" we discussed in Chapter 12 or that fills a known skill gap in your department.

Examples of High-Value Graduate Certificates:

For Mechanical/Aerospace Engineers: Look for programs like a Certificate in Composite Materials or a Certificate in Hypersonics, like the one offered by Purdue University.

For Electrical/Computer Engineers: A Graduate Certificate in RF & Microwave Engineering is a classic choice, while a Certificate in Artificial Intelligence and Machine Learning, offered by universities like Stanford or MIT, is a powerful modern credential.

For All Disciplines: A Graduate Certificate in Systems Engineering or Systems Architecture, like those offered by institutions such as Johns Hopkins University or The Stevens Institute of Technology, is one of the most respected and versatile credentials you can earn in the defense industry.

The "Spike" vs. The "Pivot"

A Graduate Certificate is not just a credential; it is a powerful tool for deliberately shaping your career. There are two primary strategic ways to use it:

The Spike: This is the most common and direct path. If you are a mechanical engineer already working in a structures group, pursuing a Graduate Certificate in Composite Materials is a "spike." You are taking your existing expertise and making it deeper, sharper, and more valuable. This is the fastest way to become a Subject Matter Expert (SME) on your team, signaling to your manager that you are committed to technical excellence in your current role. This is an excellent choice for an engineer on the Technical Fellowship path.

The Pivot: This is a more advanced and powerful strategic move. Perhaps you are a mechanical engineer who has become fascinated by the AI and autonomy work being done by the software team on your program. A Graduate Certificate in Artificial Intelligence and Machine Learning can be your bridge to that world. It provides you with the formal, foundational knowledge to have intelligent conversations with that team and demonstrates a credible, proactive interest in their discipline. This can be the key that unlocks the opportunity to take on a cross-disciplinary task, and eventually, to pivot your career into that new, high-demand field. For any engineer who wants to become more "T-shaped," the certificate is the perfect tool for building the horizontal bar.

Part 3: Forging the Habits of a Lifelong Learner

The final, and perhaps most important, pillar of your professional development is the commitment to informal, continuous learning. This is not about degrees or certifications; it is about cultivating a mindset of insatiable curiosity. This is the daily, weekly, and monthly habit of staying current, of broadening your horizons, and of learning from the experts around you. This is what truly separates the great engineers from the merely good ones, and it is a practice that you must cultivate from the very first day of your career.

1. The Power of Internal Training

Large companies offer a vast catalog of internal training courses that are a hidden goldmine of specialized, often classified, knowledge. These can range from week-long "deep dives" into a specific radar system, taught by the Technical Fellow who designed it, to short courses on new software tools or a new company process. You must take advantage of every opportunity you can. These courses are not just a chance to learn; they are a chance to network with experts from other parts of the company and to signal to your manager that you are invested in your own growth. Do not wait for your manager to assign your training; be proactive, browse the internal training catalog, and ask to be sent.

2. Building Your Professional Library

Serious professionals are serious readers. Staying on the cutting edge requires a commitment to reading beyond your immediate task list. Below are some initial recommendations but a more comprehensive list can be found at the end of this book in the Book Recommendations and Further Reading section.

The Trade Journals: Dedicate time each week to read the key industry publications. Websites like Breaking Defense, Defense News, and Aviation Week & Space Technology are the "daily papers" of the industry.

The Think Tanks: To understand the future, read the reports from major think tanks like the RAND Corporation and the Center for Strategic and International Studies (CSIS). Their analysis often precedes and shapes future government strategy.

The Foundational Books: Ask your mentors: "What is the one book on your shelf that you've referenced most in your career?" Build your physical and digital library with these foundational texts on topics from control theory to systems engineering.

3. Conferences and Publications

Attending and, eventually, presenting at technical conferences is a key part of the life of a senior engineer. As a junior engineer, ask your manager if there is a budget for you to attend a major conference in your field (like the AIAA SciTech Forum). Go there to learn, to listen to the technical papers, and to see what other companies and universities are working on. As you become more senior and develop your own expertise, your goal should be to present your own work, to contribute your own knowledge back to the community, and to build your professional reputation outside the walls of your own company.

The Lifelong Launchpad

Your undergraduate degree got you the job. It proved you have the raw intellectual horsepower and the discipline to solve difficult problems. It is the price of admission. Your commitment to continuous learning will, however, be what defines the entire trajectory of your career. With the pace of technological advancements, especially in the defense sector, you must pursue all avenues of furthering your education. With your new role, opportunities will be presented that allow you to pursue a Master's degree, Doctorates, or certificate, all on the company's dime.

The specific technologies you master today may become obsolete, but your ability to learn will not. By dedicating yourself to this lifelong pursuit of a sharper edge, you are not just building a more impressive resume; you are forging a more resilient and rewarding career. You are ensuring that you will always be ready to adapt, to contribute, and to lead in the face of the next great challenge.

Conclusion

We began this journey together on a ridge, overlooking a vast and sprawling landscape of opportunity. In the distance were the familiar paths into commercial tech, traditional industry, and academia. But there was another path, the one that led into the world of aerospace and defense, a path that felt like a "black box" shrouded in secrecy and speaking a language all its own.

You now hold the key to that box. Together, we have turned the key. We have pulled back the curtain and demystified the process. You have learned the history that shaped this unique culture and mapped the industrial ecosystem of Primes, subcontractors, and national labs. You have taken the first step into understanding the five great Arenas of Innovation, from the brutal physics of the Air Domain to the silent depths of the Sea. You have been given the strategic foresight to see the megatrends that are shaping the future battlefield and a guide to cultivating the modern skillsets you will need to thrive there.

You have learned the tactics to forge your credentials in college, to weaponize your projects, and to conquer the internship gauntlet. You have been given the playbook for the application arsenal; the verbatim scripts, templates, and "insider" strategies to build a resume and cover letter that get noticed. You have learned how to navigate the great filters of the interview and the security clearance. And finally, you have been given the unwritten rules for not just surviving, but succeeding, in your first years on the job.

The map is now complete. But a map is not the territory. The goal of this book was never just to give you information; it was to instill in you a mindset. As we part ways and you begin your own journey, there are three fundamental tenets from our time together that will serve as your compass.

First, *this is a mission, not just a job.* The single greatest differentiator between a career in this industry and any other is the profound sense of purpose; the "gravity" of the work. You are not just debugging an app or optimizing a supply chain. You are building the systems that protect a nation and its allies. You are the steward of the tools that soldiers, sailors, airmen, and guardians rely on in moments of ultimate consequence. Never lose sight of this "why." Let it be the fuel that drives your commitment to rigor, your attention to detail, and your pursuit of excellence when the work is hard and the hours are long.

Second, *trust is the only currency that matters.* Your technical skills, proven by your GPA and your projects, are the price of admission. They will get you in the door. But your career will be built on the invisible currency of trust. It is a currency you will earn in the small, consistent actions of your daily work: by revering the process, by meticulously keeping your word, by owning your mistakes completely, by communicating with clarity, and by showing respect for the expertise of those around you. Your technical competence proves you are smart; your character proves you can be trusted. The most successful and respected engineers are the ones who have an abundance of both.

Finally, *you are a lifelong apprentice to your craft.* Your education did not end on graduation day; it began. In an industry defined by a relentless technological race against intelligent adversaries, the engineer who stops learning becomes obsolete. Your most valuable asset is not what you know today, but your ability and your willingness to learn what you will need to know tomorrow. The pursuit of a sharper edge, through formal degrees, professional certifications, and a daily habit of insatiable curiosity, is not an extra task. It is the job.

Your mission, should you choose to accept it, is no longer to simply understand this world. It is to help build it.

You are no longer the student standing on the ridge, looking down at an intimidating and unknowable black box. You have journeyed through its internal architecture, learned its secret languages, and understood the principles that govern it. You now have the tools, the mindset, and the map to descend into that world, not as a visitor, but as a creator.

Your role now is to design and forge the next generation of black boxes; the complex, high-consequence systems that will define the frontiers of technology and protect the future. It is a mission that will demand more from you than any academic project ever could. It will require a higher standard of diligence, precision, and character than many other fields. It will not always be easy.

But the rewards are commensurate with the challenge. It is a career that offers the rare opportunity to work on systems that define the very edge of what is possible, to build a stable and prosperous life upon a bedrock of long-term projects, and to contribute directly to a mission of global significance.

The work is hard. The stakes are high. The mission is worthy.

A Final Word of Advice.

You are going to feel overwhelmed. In your first week, your first month, maybe even your first year, there will be moments when you are surrounded by a sea of acronyms, sitting in a meeting about a system so complex it feels unknowable. You will feel a profound sense of "imposter syndrome." You will wonder if they made a mistake in hiring you.

Know this: Every single person in that room felt that exact same way on their first day. Consider it a rite of passage.

The secret is not to know everything. The secret is to be relentlessly curious, to be humble enough to ask questions, and to have the discipline to write down the answers. Show up prepared. Be a good teammate. Take pride in your craft. And on the hard days, remember the mission.

Acknowledgements

No engineering project is a solo endeavor. The most complex and rewarding systems are the product of a team, built on a foundation of collaboration, peer review, and the shared knowledge of those who came before. This book is no exception. It is not the product of a single voice, but a collection of hard-won wisdom, generously shared by a community of dedicated professionals.

My deepest gratitude goes to the senior engineers, the Chief Engineers, the Technical Fellows, and the Program Managers who, over the course of my career, have acted as my mentors. You are the giants on whose shoulders this work stands. Thank you for your patience in answering my endless questions, for pulling back the curtain on the "unwritten rules," and for taking the time to teach a young engineer the difference between a classroom theory and a flight-proven solution. The lessons you taught in quiet moments at the whiteboard are the foundation of this book.

To the colleagues and friends who served as the first "peer review" for this manuscript, thank you for your sharp questions, candid feedback, and unwavering support. You poked holes in weak arguments, flagged unclear explanations, and helped machine rough drafts into a much stronger final product. Your diligence and willingness to challenge my assumptions made this book immeasurably better.

A special thanks to the professors and early mentors who first sparked my interest in this field and who took a chance on a young engineer. Your guidance provided the initial trajectory for the rewarding career that ultimately inspired the lessons in this book.

To my family and friends, your patience and encouragement through the countless late nights and weekends spent writing this guide were the quiet, unseen support structure that made it all possible. Thank you for believing in this project and in me.

And finally, to you, the reader. This book was written for the ambitious, curious, and dedicated student you are. Thank you for investing your time and allowing me to share this guide with you. You are the next generation of builders, problem-solvers, and innovators who will design the systems that protect our future. My deepest hope is that this book helps you build a career of purpose, integrity, and immense satisfaction.

Thank you.

Appendix

Appendix A: The "Launchpad" Resume Template

This appendix contains the "Launchpad" resume template, a two-page format designed specifically for aspiring engineers targeting the defense and aerospace industry. It is structured to be both keyword-rich for Applicant Tracking Systems (ATS) and clean/readable for human recruiters.

More than just a template, this is an annotated guide. For each section, we will provide commentary on the strategic intent and the common mistakes to avoid. Remember the core principles:

Two Pages is Standard: It gives you the necessary space to detail your projects and skills.

Keywords are Critical: Align the skills and course names with the language used in the job description.

Evidence is Everything: Use the STAR method in your bullet points to show, not just tell, what you accomplished.

[Your Name]
[City], [State] [Zip Code] | [Your Phone Number] |
[Your.Email@server.com] | [linkedin.com/in/your-custom-url]
U.S. Citizen
Security Clearance: [e.g., Active Secret, or Secret Eligibility (Investigation
Initiated Mo/Yr)]

This header is clean, professional, and contains the single most important keyword for a recruiter: 'U.S. Citizen.' We also add a dedicated line for a Security Clearance. If you have one from a prior internship, this is a top-tier credential that must be listed prominently. Putting these items at the very top is a sign of professionalism and shows you understand the industry's non-negotiable requirements.

EDUCATION
[Name of Your University] – [City], [State]
Bachelor of Science in [Your Major] – Expected [Month] [Year]

Minor in [Your Minor, if applicable]

Overall GPA: [Your GPA] / 4.0

- **Relevant Coursework:** [List 6-8 of your most relevant, impressive upper-level courses.]

> *Your education is your primary credential as a student, so it belongs at the top. The "Relevant Coursework" section is a critical area for keyword optimization. Use the Keyword Blueprint you created in Chapter 15 to select the courses that are most relevant to the specific job you are applying for. For a robotics role, list "Control Systems" and "Mechatronics." For a structures role, list "Finite Element Analysis" and "Mechanics of Materials.*

SKILLS SUMMARY

Software & Tools: [List all relevant software. Be specific. Include CAD, Analysis, Programming, etc.]

Technical Skills: [List your conceptual and theoretical skills. Mirror the job description.]

Lab & Shop Skills: [List any hands-on experience with tools and equipment.]

> *This section is your primary weapon for defeating the ATS. It should be a dense but organized block of the exact keywords from the job description. The subcategories ("Software," "Technical," "Lab") are crucial. They allow a human recruiter, who is skimming your resume in six seconds, to immediately find the information they are looking for. Do not be humble here; if you have used a piece of software or a tool even once for a class project, you can list it.*

PROJECTS

[Title of Your Capstone or Major Project] | [City], [State]

Senior Capstone Design Project – [Start Date] – [End Date]

- [Action Verb] [Describe your specific task or objective]. [Action Verb] [Describe the specific, individual action you took using relevant tools or skills]. [Result] [Quantify the outcome of your action].
- [Action Verb]... [Quantifiable Result]...
- [Action Verb]... [Quantifiable Result]...

This is the heart of your resume. It is your evidence locker. Notice the structure: a bold, descriptive title, your affiliation, and then a series of powerful, quantifiable bullet points. As we discussed in the masterclass in Chapter 16, every bullet point must be a mini-story of accomplishment, starting with a strong action verb and ending with a number or a metric that proves your impact. This is what separates a professional resume from an amateur one.

(*Page 2 Begins Here*)

PROFESSIONAL EXPERIENCE

[Name of Company] – [City], [State]

[Your Title, e.g., Mechanical Engineering Intern] – [Start Date] – [End Date]

[Use the exact same STAR-method, quantifiable bullet point format as your Projects section.]

[Describe your accomplishments, not your duties.]

> *The "Professional Experience" section uses the exact same format as "Projects" to maintain a clean, consistent, and professional look. If you have had a relevant internship, this section should come before your Projects section. If your projects are your strongest experience, they should come first. Always lead with your most impressive evidence.*

LEADERSHIP & ACTIVITIES

[Name of Club or Organization, e.g., American Institute of Aeronautics and Astronautics (AIAA)]

[Your Title, e.g., Team Lead, Design/Build/Fly Competition] – [Start Date] – [End Date]

[Use a bullet point or two to describe your role or accomplishments, quantifying them if possible (e.g., "managed a $5,000 budget" or "increased membership by 20%").]

> *This section is your proof of "soft skills." It shows that you are not just a technically competent engineer, but also a leader and a team player. A leadership role in a technical club is far more valuable than a passive membership. It provides a*

source of powerful stories about teamwork and project management for your interviews.

Action Verb Arsenal

A weak resume is filled with passive phrases like "was responsible for" or "assisted with." A powerful resume starts every bullet point with a strong, active verb that shows you are a doer. Use this arsenal of verbs to craft more compelling and professional bullet points for your "Projects" and "Experience" sections.

For Leadership & Management:

Led

Managed

Coordinated

Directed

Organized

Planned

Oversaw

Supervised

For Design & Creation:

Designed

Developed

Created

Built

Fabricated

Implemented

Modeled

Prototyped

For Analysis & Problem-Solving:

Analyzed

Evaluated

Calculated

Investigated

Troubleshot

Diagnosed

Quantified

Validated

For Improvement & Optimization:

Improved

Optimized

Increased

Reduced

Streamlined

Enhanced

Refined

Upgraded

For Communication & Collaboration:

Presented

Authored

Documented

Collaborated

Proposed

Instructed

Final Note for the Reader: After filling out this template, always save and submit your resume as a PDF file to preserve formatting and appear professional. Name the file clearly: FirstName_LastName_Resume.pdf. This template is a guide; if your experience warrants more than two pages of rich, relevant, and quantifiable accomplishments, do not be afraid to use the space you need.

Appendix B: The Interview "Mission Prep" & Post-Mission Follow-Up Sheet

An interview is a mission. Success is not a matter of luck; it is the direct result of deliberate preparation, intelligence gathering, and having a clear plan. Do not walk in unprepared. This worksheet is your mission preparation and follow-up tool. It is designed to transform your interview prep from a passive activity (reading and worrying) into an active one (writing and strategizing).

The physical act of writing down your thoughts will solidify them in your mind, organize your narrative, and give you a powerful sense of confidence. Print out a fresh copy for every single interview you have. Complete *Sections 1-4* the night before your interview to get your mind focused. Review it again one hour before the meeting to get "in the zone." Complete *Section 5* immediately after your interview while the details are still fresh in your mind.

Interview Mission Sheet

MISSION OVERVIEW

(This is your logistical checklist. Getting the details right is the first sign of a professional.)

Company: _____

Position: _____

Date & Time: _____

Interviewer(s) & Title(s): _____

1. INTELLIGENCE GATHERING (Company & Role Research)

Company's Primary Mission/Product at this Location:

(e.g., "This Lockheed Martin facility in Orlando focuses on missile systems and sensor design.")

_____

_____

One Specific Program or Technology They Work On That Interests Me (Your Hook):

(e.g., "The Long Range Anti-Ship Missile (LRASM) program.")

_____

_____

The Top 3 "Preferred Skills" Listed in the Job Description (Your Keywords):

_____

_____

_____

2. MY "ELEVATOR PITCH"

(Structure: "I am a [Year] [Major] at [University] with a focus on [Specialty]. My experience with [Project X] gave me strong skills in [Skill 1] and [Skill 2], and I am excited about this role because [Reason].")

_____

_____

_____

_____

_____

_____

3. MY STAR STORIES

Story #1: A Major Technical Challenge

S (Situation): _____

T (Task): _____

A (Action): _____

R (Result): _____

Story #2: A Teamwork / Leadership Example

S (Situation): _____

T (Task): _____

A (Action): _____

R (Result): _____

Story #3: A Failure or a Time I Learned Something New

S (Situation): _____

T (Task): _____

A (Action): _____

R (Result): _____

4. MY QUESTIONS FOR THEM

About the Role:

(e.g., "What does a typical day look like for an intern on this team?" or "What is the single biggest challenge the team is facing right now that an intern could help with?")

_____

_____

About the Team/Culture:

(e.g., "How does your team handle knowledge sharing and mentorship for new engineers?")

_____

_____

About the Future:

(e.g., "What are some of the key technologies or skills that you see becoming most important for your group over the next few years?")

_____

_____

About the Interviewer (If appropriate):

(e.g., "What has been the most interesting or rewarding project you've worked on here at [Company Name]?")

_____

_____

POST-MISSION: THE FOLLOW-UP (Within 24 Hours)

Objective: To thank the interviewer(s) for their time, briefly reiterate your interest, and reference a specific point from your conversation to prove you were listening.

Instructions: Within 24 hours of your interview, send a concise, professional email to the main interviewer(s). Use the template below.

Email Address(es) of Interviewer(s): _____

Key Conversation Point to Reference (Your Proof of Listening):
(What was one specific, interesting thing you discussed? A project they mentioned? A challenge they described? This is what elevates your email from a generic template to a personal note.)
(e.g., "I particularly enjoyed our discussion about the challenges of thermal management for the new sensor pod.")

_____

_____

_____

_____

Follow-Up Email Template

Subject: Thank You - Interview for [Position Title]

Dear [Mr./Ms. Last Name of Interviewer],

Thank you again for taking the time to speak with me today about the [Position Title] role. I truly enjoyed our conversation and learning more about the important work your team is doing at [Company Name].

I was particularly interested in our discussion about [Reference the Key Conversation Point you wrote down above]. It solidified my enthusiasm for this opportunity and my belief that my skills in [Mention 1-2 key skills, e.g., "thermal analysis and systems engineering"] would allow me to be a valuable contributor to your team.

I am very excited about the possibility of joining [Company Name]. Thank you again for your time and consideration.

P.S. I particularly enjoyed hearing about [mention a personal, non-work detail they shared, e.g., your experience hiking in the Rockies].

Sincerely,

[Your Name]

[Your Phone Number]

[Link to your LinkedIn Profile]

PERSONAL DEBRIEF (To Be Completed Immediately After the Interview)

What was the one question that I answered BEST? Why did it work?

_____

_____

What was the one question that I struggled with the MOST? How can I prepare a better answer for next time?

_____

_____

What technical concepts came up that I felt rusty on? What should I review before my next interview?

_____

_____

Did I learn anything new or surprising about the company, the culture, or the role?

_____

_____

Did I "click" with the interviewer? What did I learn about their personality and what they value (e.g., directness, technical depth, humor)?

_____

_____

Appendix C: The Career Fair "Mission" Checklist

The career fair is a high-speed, high-stakes environment. It is a chaotic mix of long lines, loud crowds, and brief, high-pressure conversations. It is very easy to become overwhelmed, to waste time, and to walk away feeling like you missed your chance. A successful outcome is not a matter of luck; it is the direct result of deliberate preparation and a clear plan of execution.

This checklist is designed to be a simple, actionable tool to ensure you are fully prepared to make the most of this critical event. Print this out a week before the fair and use it to track your progress. Walking into that hall with a plan is what will give you the confidence to stand out from the crowd.

Phase 1: T-Minus 1 Week

This is where you win the battle before it even starts. The work you do in the week leading up to the fair will determine 80% of your success. Preparation is a form of respect, and it is immediately obvious to recruiters.

[ ] Obtain List of Attending Companies: Go to your university's career services website and download the official list and map of employers who will be at the fair.

[ ] Create a Tiered Target List: Do not wander aimlessly. Create a strategic plan of attack.

[ ] Identify your 3-5 "Tier 1" companies. These are your top priorities, your dream internships.

[ ] Identify your 5-10 "Tier 2" companies. These are your secondary targets, excellent companies that you are also highly interested in.

[ ] Note the location of your Tier 1 and Tier 2 companies on the map.

[ ] Visit the "Careers" page of each Tier 1 company to see what specific internship positions they have open.

[ ] Identify one or two specific programs (e.g., "the F-35 program"), technologies, or local site functions (e.g., "your radar division in

Baltimore") that genuinely interest you for each company. This is the raw material for your elevator pitch.

[ ] Finalize and Polish Your Master Resume: Update your master resume with your latest GPA, projects, and skills. Proofread it for typos at least three times.

[ ] Create a Tailored Resume for the Fair: Create a slightly tailored version of your resume that emphasizes the skills and coursework most relevant to your Tier 1 companies.

[ ] Write out your 30-second introduction using the formula from Chapter 19.

[ ] Practice saying it out loud in front of a mirror or record yourself on your phone until it sounds natural, confident, and conversational, not robotic.

[ ] Pre-Apply Online to Your Tier 1 Targets: Many large companies will ask, "Have you already applied online?" Being able to say "Yes, I applied for requisition #12345" shows significant preparation and makes it easier for them to track you. Do this a few days before the fair.

Phase 2: T-Minus 1 Day

The night before is about eliminating all sources of day-of stress. Your only job on mission day should be to execute your plan, not to worry about a wrinkled shirt or a forgotten resume.

[ ] Print High-Quality Resumes: Print at least 20 copies of your tailored resume on dedicated, high-quality resume paper.

[ ] Place your resumes in a clean, professional portfolio or folder.

[ ] Include a quality pen and a small notepad for taking notes.

[ ] Include a copy of your target list and the career fair map.

[ ] Lay out your full professional outfit (a suit or equivalent is the standard).

[ ] Ensure clothes are clean, ironed, and ready to go. Polish your shoes.

[ ] Pack Your Bag: Include your folder, breath mints (a must), and a bottle of water.

[ ] Get a Good Night's Sleep: A rested, alert, and engaged mind is your single best asset.

Phase 3: The Day of the Interview

This is game time. Your preparation allows you to be calm, confident, and strategic in a chaotic environment.

[ ] Eat a Good Meal Beforehand: You will be on your feet for hours. You'll need the energy.

[ ] Arrive Early: Get there 15-20 minutes before the doors officially open. This allows you to get a feel for the layout and be one of the first in line at a top-tier company before the massive crowds arrive.

[ ] Navigate the Floor: The "Line" vs. "The Crowd"

> *The Line: This is straightforward. While you are waiting, use the time productively. Do not stare at your phone. Listen to the conversations the students in front of you are having with the recruiter. This is a live intelligence-gathering opportunity to learn what they're interested in. Rehearse your own pitch one last time in your head.*
>
> *The Crowd: This is more challenging. At the most popular booths, there may be a chaotic "crowd" instead of a line. Waiting politely on the edge is a strategy for failure. You must be professionally assertive.*
>
> *Work your way respectfully to the front of the crowd, near the table. Do not interrupt an ongoing conversation.*
>
> *As one conversation is ending, try to make direct eye contact with the recruiter. A simple, confident look is often all it takes for them to acknowledge you as the next person.*
>
> *When you have your opening, be ready. Offer your hand for a handshake and immediately launch into your well-practiced elevator pitch.*

[ ] Execute a "Warm-Up" Conversation: Do not go to your number one dream company first. Visit one of your Tier 2 targets to shake off the nerves and practice your pitch in a lower-stakes environment.

[ ] Execute on Tier 1 Targets: Approach your top-priority companies with confidence.

[ ] Make a Professional First Impression: When it's your turn, ignore the chaos around you. Make direct eye contact, offer a firm handshake, and deliver your well-practiced elevator pitch.

[ ] Take Notes After Each Key Conversation: After Each Key Conversation: As soon as you walk away from a booth, step to the side and jot down three things on your notepad: (1) The recruiter's name, (2) One specific, memorable technical or personal detail from your conversation

(e.g., "discussed their work on the GPS III program," or "also a fan of hiking"), and (3) Their key piece of advice (e.g., "apply online to req #12345"). This is not optional; it is critical for your 24-hour follow-up.

[ ] Collect Business Cards: If possible, ask for a business card. If they don't have one, ask for the best way to contact them and write down their name and email address.

[ ] Execute the 24-Hour Follow-Up: Within 24 hours of the fair, send a personalized thank you email or LinkedIn connection request to the key people you spoke with, referencing the specific notes you took after your conversation. This is what separates the professional from the amateur.

Post-Mission Success: A career fair is not won at the event itself; it is won in the preparation and the follow-up. By following this plan, you have done more than just hand out resumes. You have gathered intelligence, made targeted human connections, and demonstrated a level of professional maturity that will make you stand out from the crowd.

Appendix D: The Internship Pre-Flight Checklist

You've done the hard work: you aced the interview and landed the internship. Congratulations! Now, the next mission begins: making a powerful first impression from the moment you walk through the door. A successful first week is not a matter of luck; it is the direct result of careful and deliberate preparation in the weeks leading up to your start date. Feeling prepared will not only reduce the natural and completely normal "first-day jitters," but it will also allow you to focus on what really matters, learning your new role and connecting with your new team.

This checklist is your pre-flight guide. It is designed to be a simple, actionable tool to ensure you have all of your logistical, professional, and personal ducks in a row before Day 1. Think of this as the formal checklist a pilot goes through before takeoff. By methodically working through these steps, you can walk into the facility on your first day feeling calm, prepared, and confident, ready to begin the real mission: crushing your internship.

Phase 1: T-Minus 1 Month

This is the phase for handling the big-picture logistics, the major sources of stress that can derail your focus if left to the last minute. Getting these sorted out early is a sign of professional maturity and will give you immense peace of mind.

[ ] Security Paperwork Submitted: This is your number one priority. You will receive instructions from the company's security office to fill out your security clearance paperwork (the SF-86 via the e-QIP portal). Do not procrastinate on this. Complete it as quickly and as accurately as possible. The sooner you submit, the sooner your investigation can begin, which is a critical step for your internship. Double-check every entry before you hit "submit."

[ ] Housing Finalized: If your internship is in a different city, you must secure your summer housing now. Do not wait until the last minute and hope for the best. Whether it's a university sublet, a corporate housing option provided by the company, or a short-term apartment lease, having a signed agreement and a confirmed move-in date is a critical logistical step that removes a massive variable from your plate.

[ ] Transportation Plan Established: How will you get to the facility every day? Research your commute. Use Google Maps to check the route during rush hour traffic to see how long it *really* takes. If you are relying on public transportation, map out the route and the schedule. Do a test drive or a test bus ride a week or two before you start. Being late in your first week is not an option.

[ ] Professional Wardrobe Acquired: The standard dress code for most defense contractors is "business casual." This is not the time for t-shirts

and ripped jeans, but you also don't need a full suit every day. Invest in a solid, professional, and comfortable wardrobe:

-Several pairs of chinos or slacks (e.g., navy, khaki, gray).
-A collection of professional, collared shirts (polo shirts or button-downs are perfect).
-A pair of clean, professional shoes (not sneakers).
-Have at least one full business professional outfit (a suit or equivalent) clean and ready for any potential high-level meetings or your final presentation.

[ ] Confirm HR Onboarding Details: Read through your official offer letter and any onboarding emails from Human Resources with the attention of an engineer reviewing a requirements document. Add all critical dates, times, required documents, and specific locations to your personal calendar.

Phase 2: T-Minus 1 Week

The logistics are handled. This week is about preparing yourself professionally and mentally. It's about sharpening your intellectual sword so you can make a great first impression in your initial conversations.

[ ] Company Research Refresh: You did your research to get the interview; now do it again, but with the mindset of an employee, not a candidate. Go to the company's website and the websites of news outlets like "Breaking Defense." What has happened in the last few months? Did they have a major program success or win a new contract? Being current on the company's activities shows you are engaged and interested beyond just your own role.

[ ] Make the LinkedIn Connection: Find your future manager and your assigned mentor (if you know who it is) on LinkedIn. Send them a brief, professional connection request.

> *Example Script: "Hi [Manager's Name], I'm [Your Name], and I'm very much looking forward to starting my internship with your team on [Date]. Just wanted to connect here beforehand. Best, [Your Name]."*
> *This is a simple, professional touch that gets your name in front of them, shows initiative, and demonstrates your enthusiasm.*

[ ] Review Your Technical Fundamentals: You are not expected to be an expert on day one, but you are expected to remember your core engineering knowledge. Spend a few hours reviewing the coursework that is most relevant to the job description. If you are going into a structures group, re-read your statics and mechanics of materials notes. If you are going into an RF group, refresh your memory on Maxwell's equations. This will help you feel more confident and less intimidated in your first technical conversations with senior engineers.

[ ] Purchase Your Notebook: Go to an office supply store and buy a high-quality, professional notebook (a simple, black, hardcover notebook is perfect) and a good pen that you enjoy writing with. As we discussed

in the "First 90 Days" chapter, this will be your single most important tool in your first week for capturing the flood of new information.

[ ] Memorize the Names of Your 'Day One' Contacts: You will have the names of your manager and likely your assigned mentor. Commit them to memory. In the whirlwind of your first day, being able to walk up and say, "Hi Sarah, my name is Alex, it's a pleasure to meet you" with confidence, rather than fumbling for your phone to look up their name, makes a powerful and respectful first impression.

Phase 3: T-Minus 1 Day

This is about setting yourself up for a smooth, stress-free first morning. Your goal is to eliminate any and all sources of potential chaos so you can walk in feeling calm, collected, and ready to focus.

[ ] Lay Out Your Professional Outfit: Have your entire outfit for the next day (ironed shirt, pants, shoes, socks, belt) laid out and ready to go. This eliminates any morning decision-making and the risk of a last-minute wardrobe malfunction.

[ ] Pack Your Bag: Prepare the bag you will take to work the night before, not the morning of. It should include:

> *-Your new mission notebook and several pens.*
> *-Any required identification (Driver's License, Social Security Card, Passport) as specified by HR for your Day 1 badging process.*
> *-A copy of your offer letter and any onboarding instructions with the address and the name of the person you are supposed to meet.*
> *-Your lunch or lunch money.*
> *-A bottle of water and a small, quiet snack (like a granola bar).*

[ ] Confirm Your Route and Wake-Up Time: Double-check your commute plan and your morning timeline. Know exactly when you need to wake up and when you need to leave the house to arrive 15 minutes early.

[ ] Set Two Alarms: One on your phone, one on a separate, battery-powered alarm clock. There is absolutely no excuse for being late on your first day.

[ ] Get a Good Night's Sleep: This is not a night for an all-nighter. Do not stay up until 2 AM playing video games or scrolling through social media. A rested, alert, and engaged mind is your best asset for absorbing the massive amount of new information you will receive and for making a great first impression.

By methodically working through this checklist, you have eliminated the variables and reduced the stress of your first day. You can walk into the facility feeling calm, prepared, and confident, ready to focus on what truly matters: learning your new role, connecting with your new team, and executing your real mission: crushing your internship.

Appendix E: A Guide to the Security Briefing and the Classified World

Welcome to the Vault

It happens about a week or two into your new job. Your manager, "Sarah," stops by your desk. "Great news," she says, "your interim clearance was just granted. Let's go see the FSO to get you read in." This is the moment you've been waiting for, the official start of your real work.

You walk with Sarah to the security office, a quiet and professional part of the building. The Facility Security Officer (FSO), a serious but helpful professional named "Mark," invites you into his office. He doesn't interrogate you; he congratulates you. He calmly and methodically explains your responsibilities as a cleared individual. He talks about protecting information, about reporting requirements, and about the trust that the government has placed in you.

Then, he places a document in front of you. It's the *Standard Form 312 (SF-312), the Classified Information Nondisclosure Agreement.* It is a legally binding contract between you and the United States government. As you sign and date it, you feel the tangible weight of the moment. This is a profound commitment, a promise that you will protect the nation's secrets for the rest of your life.

Later that day, your mentor, "Dave," walks you to a heavy, unmarked door with a prominent cipher lock on the wall next to it. "Ready to see where the real work happens?" he asks with a smile. He expertly punches in a code, and the lock buzzes open with a solid "thunk." He pulls the heavy door open, and you step through. The door closes behind you with a "whoosh," and the noise of the outside office is instantly gone.

You are now in the "vault," the SCIF (Sensitive Compartmented Information Facility). There are no windows. The air is quiet and still. The desks are clean, and the documents on them are covered with brightly colored sheets: red for SECRET, blue for CONFIDENTIAL. Dave sees the look of awe and intimidation on your face. "It feels like a different world in here, doesn't it?" he says. "Don't worry, you'll get used to it. The most important rule here is simple: if you don't know, ask. We're all on the same team when it comes to protecting the work."

This appendix is your guide to this new world. It is designed to demystify the day-to-day reality of working with classified information and to give you the confidence to operate professionally and securely from your very first day.

"Reading In": Your First Briefing and the Non-Disclosure Agreement

Your first official step into the classified world will be a formal meeting with your company's security officer. This is often referred to as an "indoctrination" or, more commonly, a "read-in." This is not a test; it is a formal briefing and a contractual agreement.

The security officer will sit down with you one-on-one. They will explain the specific security protocols for your program and your facility. They will detail your responsibilities as a cleared individual. At the end of this briefing, you will be asked to sign a *Standard Form 312 (SF-312), the Classified Information Nondisclosure Agreement.*

The SF-312 is a legally binding contract between you and the United States government. By signing it, you are acknowledging that you understand your responsibilities and that you are agreeing, for the rest of your life, to never disclose the classified information you are about to receive to any unauthorized person. It is a moment of profound gravity and is the formal entry point into this community of trust.

A Guide to The SCIF

For many engineers, especially those working on Top Secret or SCI programs, a significant portion of the workday will be spent inside a SCIF (Sensitive Compartmented Information Facility). A SCIF is not just a regular conference room with a strong lock. It is a secure vault, an architectural and electronic fortress designed to prevent any information from leaking out. Understanding the rules of this environment is non-negotiable.

No Personal Electronics, Ever. This is the most important and absolute rule. You are not allowed to bring any personal electronic devices (cell phones, smartwatches, fitness trackers, personal laptops, wireless headphones) into a SCIF. This is because any device with a microphone, camera, or wireless transmission capability represents an unacceptable risk for surreptitious listening or data exfiltration. There is zero tolerance for this rule. Most facilities will have small lockers outside the entrance where you can store your personal items.

Secure Containers ("Safes"). All classified hard drives, documents, and other materials are stored in heavy, GSA-approved security containers, which look like large, gray safes. You will be taught the procedure for opening and, most importantly, properly securing these containers. This involves spinning the combination lock multiple times to ensure it is fully reset.

"Clean Desk" Policy. At the end of every single workday, no classified material can be left out on a desk or a workbench. Every single document, every notebook, and every hard drive must be returned to its proper secure container. A supervisor or security officer will often perform a walk-through at the end of the day to ensure the "clean desk" policy has been followed.

Controlling the Door. The door to a SCIF is a heavy, vault-like door with a cipher lock. You must never "tailgate" someone into a SCIF (follow them in without entering your own code), and you must never hold the door open for someone you do not know and cannot verify is authorized to be there.

Handling Classified Material: A Practical Guide

Handling your first classified document can be a nerve-wracking experience. The procedures are simple, but they must be followed with meticulous discipline.

The Cover Sheets: You will learn the visual language of the colored cover sheets that are required to be on the front of any classified document when it is not in a safe. These are not just pieces of paper; they are legal warning barriers.

Purple (CUI): Standard Form 901. This stands for Controlled Unclassified Information. It has largely replaced the older "FOUO" (For Official Use Only) designation. While not "classified" in the national security sense, CUI documents (like designs or test plans) must still be covered when on your desk and locked away at night.

Blue (CONFIDENTIAL): Standard Form 705.

Red (SECRET): Standard Form 704.

Orange (TOP SECRET): Standard Form 703.

Yellow (TOP SECRET / SCI): While there isn't one single universal "Standard Form" number for SCI documents, facilities often use bright yellow or yellow-striped cover sheets to denote Sensitive Compartmented Information, ensuring it is never accidentally removed from a SCIF.

These cover sheets are a constant, visual reminder of the sensitivity of the information you are handling.

Markings and Portion Marking: Every classified document is covered in specific markings. You will receive training on this, but you should know the concept of "Portion Marking." Every single paragraph, bullet point, and title in a document will be preceded by a marking in parentheses, like (U), (S), or (TS//NF). This tells you the classification level of that specific sentence. This allows engineers to extract unclassified data from a classified document without creating a spill.

Transportation and Destruction: You cannot simply walk down the hall with a classified document. Moving classified material, even within the same building, often requires it to be in a specific, marked folder and may require a two-person escort. Destroying classified material is also a formal process, requiring the use of a government-approved, high-security shredder or a burn bag.

GSA Container: At the end of the day, you will secure your work in a heavy, gray GSA-approved security container (often a Class 5 or Class 6 cabinet). You will likely learn to use a mechanical dial, which uses internal power generation (you have to spin the dial rapidly to "power up" the lock before entering your code).

If You Don't Know, Ask

The rules of the classified world are complex, and you are not expected to know everything on day one. The single most important rule to remember is this: *When in doubt, ask.*

The security professionals at your company, your Facility Security Officer (FSO), are not there to "catch" you making a mistake. They are your partners. Their job is to help you understand and follow the rules to protect both you and the information. It is always, always better to ask a "dumb question" about a security procedure than to make a guess and get it wrong. A simple question might feel momentarily embarrassing; a security violation can be a career-ending event. The best engineers are the ones who are not afraid to admit when they don't know something and to ask for guidance.

Glossary of Acronyms and Terms

Welcome to the "alphabet soup." The defense and aerospace industry runs on a dense language of acronyms, technical terms, and professional slang. To a newcomer, this can feel like an intimidating barrier.

This glossary is your quick-reference guide, your personal decoder for this complex world. It contains not only the formal government and engineering acronyms you will see in documents, but also the common, informal terms you will hear in the hallways and in team meetings. Use this not as a list to be memorized, but as a dictionary to be referenced whenever you encounter a term that is new or unfamiliar.

Term / Acronym	Stands For / Definition
A	
Action Items	The specific, assigned tasks resulting from a meeting, with a designated owner and a due date.
Adjudication	The final step in the clearance process where an official decides to grant or deny a security clearance.
AESA	Active Electronically Scanned Array
AFRL	Air Force Research Laboratory
AIAA	American Institute of Aeronautics and Astronautics
Anechoic Chamber	A specialized room designed to completely absorb reflections of sound or electromagnetic waves.
ARPANET	Advanced Research Projects Agency Network (the precursor to the Internet)
As-Built	The actual, physical state of a component after manufacturing, with all its variations.
As-Designed	The perfect, idealized state of a component as represented in its CAD model and drawing.
ASME	American Society of Mechanical Engineers

ATO	Authority to Operate; The formal declaration by a designated authority that an information system is approved to operate in a particular security mode using a prescribed set of safeguards.
ATS	Applicant Tracking System
AWACS	Airborne Warning and Control System

B

Baseline	The formally approved, configuration-controlled version of a design or document.
Blue Force Tracing	A system that displays the location of friendly military forces on a digital map.
Blue Team	A team designated to act as the defenders in a test or cybersecurity exercise, responsible for protecting the system against the "Red Team's" attacks.
BLUF	Bottom Line Up Front; A communication style, originating in the military, where the conclusion and recommendation are placed at the very beginning of a document or brief, followed by the supporting data.
BOM	Bill of Materials
Bunny Suit	Slang for the head-to-toe, anti-static garment worn by engineers and technicians working in a "clean room" to prevent contamination of sensitive hardware, especially for space systems.

C

C4ISR	Command, Control, Communications, Computers, Intelligence, Surveillance, and Reconnaissance
CAD	Computer-Aided Design
CAE	Computer-Aided Engineering
CAM	Computer-Aided Manufacturing
CCA	Collaborative Combat Aircraft ("Loyal Wingman")
CDR	Critical Design Review

CFD	Computational Fluid Dynamics
Chaff	Clouds of small, reflective fibers dispensed by aircraft to confuse radar.
Clean Room	A controlled manufacturing or assembly environment with a very low level of pollutants such as dust, airborne microbes, and chemical vapors. Essential for building satellite and optical systems.
CM	Configuration Management
COMINT	Communications Intelligence; A sub-discipline of SIGINT, specifically the intelligence gathered from the interception of foreign communications between people.
CONOPS	Concept of Operations
COTS	Commercial Off-the-Shelf
CSEP	Certified Systems Engineering Professional
CSWP	Certified SOLIDWORKS Professional
CYBERCOM	United States Cyber Command

D

DARPA	Defense Advanced Research Projects Agency
DCO	Defensive Cyber Operations
DCSA	Defense Counterintelligence and Security Agency
DE	Directed Energy
DEVCOM	Combat Capabilities Development Command (U.S. Army)
DFM	Design for Manufacturability
Digital Twin	A high-fidelity, living digital model of a physical system, updated with real-world operational and maintenance data, that serves as the "authoritative source of truth" for the life of a program.
DoD	Department of Defense

DoE	Department of Energy
DOORS	Dynamic Object-Oriented Requirements System
DRFM	Digital Radio Frequency Memory
DSP	Digital Signal Processing
E	
EA	Electronic Attack
ECR	Engineering Change Request
EIT	Engineer in Training; A professional designation from a state licensing board for a person who has passed the Fundamentals of Engineering (FE) exam. Often a prerequisite for becoming a Professional Engineer (PE).
ELINT	Electronic Intelligence; A sub-discipline of SIGINT, specifically the intelligence gathered from foreign non-communication electromagnetic radiations (e.g., radar signals).
EMD	Engineering and Manufacturing Development
EMP	Electromagnetic Pulse
Energetics	The field of science dealing with high-energy materials like explosives and propellants.
EO/IR	Electro-Optical / Infrared
EP	Electronic Protection
ES	Electronic Support
EW	Electronic Warfare
F	
FADEC	Full Authority Digital Engine Control
FAI	First Article Inspection
Falsification	The act of deliberately lying or omitting information on a security form like the SF-86.

FFRDC	Federally Funded Research and Development Center
Firmware	Software that is permanently programmed onto a specific piece of hardware (e.g., a microchip).
FMEA	Failure Mode and Effects Analysis
FRB	Failure Review Board
Frequency Hopping	A technique where a radio rapidly switches frequencies to avoid jamming.
FSO	Facility Security Officer
FVEY	Five Eyes (Intelligence Alliance: US, UK, Canada, Australia, NZ)

G

GaN	Gallium Nitride (a semiconductor material)
GD&T	Geometric Dimensioning and Tolerancing
GEO	Geosynchronous Equatorial Orbit; A high Earth orbit approximately 22,236 miles (35,786 km) above Earth's equator that allows a satellite to match Earth's rotation, making it appear stationary in the sky. Ideal for communications and weather satellites.
GNC	Guidance, Navigation, and Control
GOTS	Government Off-the-Shelf
GPS	Global Positioning System
GVSC	Ground Vehicle Systems Center (U.S. Army)

H

HEL	High-Energy Laser
Honeypot	A decoy computer system, set up to attract and trap cyber attackers and to study their methods.
HW	Hardware
Hypersonics	Flight at speeds of Mach 5 or higher.

I

ICD	Interface Control Document
ICWG	Interface Control Working Group
IED	Improvised Explosive Device
IEEE	Institute of Electrical and Electronics Engineers
IMU	Inertial Measurement Unit
INCOSE	International Council on Systems Engineering
Interim Clearance	A temporary security clearance granted while the full investigation is pending.
IPT	Integrated Product Team
IRAD	Internal Research and Development
ISR	Intelligence, Surveillance, and Reconnaissance
ITAR	International Traffic in Arms Regulations
IVAS	Integrated Visual Augmentation System

J

JADC2	Joint All-Domain Command and Control
Jamming	Transmitting powerful radio signals to disrupt or block an enemy's communications or radar.

K

Kill Web	A networked system where any sensor can provide targeting data to any weapon.
KPP	Key Performance Parameter; A critical attribute or characteristic of a system that is considered essential to achieving a successful mission. KPPs are formally tracked at the highest levels of a program.

L

Left of Boom	The phase of a conflict before shots are fired; the period of deterrence and disruption.
LEO	Low Earth Orbit; An orbit with an altitude of 1,200 miles (2,000 km) or less. Satellites in LEO travel at very high speeds. Ideal for Earth observation, reconnaissance, and large communications constellations like Starlink.
LIDAR	Light Detection and Ranging
LLNL	Lawrence Livermore National Laboratory
Long Pole in the Tent	Slang for the single biggest technical or schedule risk on a program, the one item that is pacing the entire effort.
LPI	Low Probability of Intercept

M

Margin	The engineering "safety buffer" in a design.
Matrixed Organization	A corporate structure where an employee reports to both a functional manager and a program manager.
MBSE	Model-Based Systems Engineering
MEO	Medium Earth Orbit: An orbit with an altitude between 1,200 and 22,000 miles (2,000 and 36,000 km). This is home to the Van Allen Belts but also the GPS constellation.
Mitigation	Steps taken to reduce the severity or likelihood of a risk.
MRAP	Mine-Resistant Ambush Protected
MRB	Material Review Board
MUM-T	Manned-Unmanned Teaming
Murder Board	Slang for a difficult, internal practice session designed to prepare a team for a formal design review.

N

NASA	National Aeronautics and Space Administration

Need-to-Know	The fundamental security principle restricting access to classified information.
NNSA	National Nuclear Security Administration
Non-Conformance	A formal term for a part that is out of tolerance and does not meet the drawing's specifications.
NRO	National Reconnaissance Office
NSA	National Security Agency

O

OCO	Offensive Cyber Operations
Offset Strategy	A high-level defense strategy that uses technological superiority to offset an adversary's numerical advantage.
OPIR	Overhead Persistent Infrared
OPSEC	Operations Security
ORNL	Oak Ridge National Laboratory

P

PDR	Preliminary Design Review
Peer Review	The process of having your design or analysis checked by your teammates for errors.
Penetration Testing	The practice of "ethical hacking" to find vulnerabilities in a computer system.
PGM	Precision-Guided Munition
Phantom Works	The official alias for Boeing's advanced development and prototyping division.
PLM	Product Lifecycle Management
PM	Program Manager
PMP	Project Management Professional

Polygraph	A lie detector test, sometimes used in high-level clearance investigations.
Prime Contractor	The main company responsible for a large defense contract.

Q

Q Clearance	A Department of Energy security clearance, equivalent to a DoD Top Secret.

R

Rad-Hard	Radiation Hardened (describes electronics designed to work in space).
RAM	Radar Absorbent Material
RCA	Root Cause Analysis
Red Line	A physical or digital, red-penned markup of an engineering drawing indicating a change.
Red Team	A team designated to act as the adversary in a test or cybersecurity exercise.
Requirements	The specific, verifiable "shall statements" that define what a system must do.
Resilience	The ability of a system or network to continue functioning even after being damaged.
RF	Radio Frequency
RFP	Request for Proposal
Right of Boom	The phase of a conflict after the shooting has started.
Risk Matrix	A chart used to manage program risks, plotting likelihood against consequence.
RTOS	Real-Time Operating System

S

SAP	Special Access Program ("black program")

SATCOM	Satellite Communications
SCI	Sensitive Compartmented Information
SCIF	Sensitive Compartmented Information Facility
Scramjet	Supersonic Combustion Ramjet, a type of engine for hypersonic flight.
SDR	Software Defined Radio
SF-86	Standard Form 86 (the security clearance questionnaire)
SIGINT	Signals Intelligence
SIL	System Integration Laboratory
Skunk Works®	The official alias for Lockheed Martin's Advanced Development Programs.
SLAM	Simultaneous Localization and Mapping
SME	Subject Matter Expert
SOW	Statement of Work
Stealth	Technology designed to make a vehicle difficult to detect by radar, infrared, or other means.
Subcontractor	A company hired by a Prime Contractor to provide a specific component.
Sustainment	The decades-long effort of maintaining and upgrading systems already in the field.
SW	Software
SWaP-C	Size, Weight, and Power - Cost
SysML	Systems Modeling Language

T

| T-Shaped Engineer | An engineer who has deep expertise in one specific discipline (the vertical bar of the 'T') but also a broad, functional |

knowledge of many other adjacent disciplines (the horizontal bar).

The -ilities	Slang for non-functional requirements (Reliability, Maintainability, etc.).
TIM	Technical Interchange Meeting
TPM	Technical Performance Measure
Trade Space	The set of all possible design solutions for a problem.
Trade Study	A formal, documented process for comparing different design options.
Traveler	A packet of paperwork (or its digital equivalent) that travels with a piece of hardware through the entire manufacturing and inspection process, containing its drawings, instructions, and a complete, auditable history of sign-offs.
TRL	Technology Readiness Level
TRR	Test Readiness Review
TS	Top Secret
TVAC	Thermal Vacuum (Chamber); A specialized test chamber used for space systems that can simulate both the hard vacuum and the extreme hot and cold temperatures of orbit.

U

UARC	University Affiliated Research Center
UAV	Unmanned Aerial Vehicle
USV	Unmanned Surface Vehicle
UUV	Unmanned Undersea Vehicle

V

V&V	Verification and Validation

W

WBS	Work Breakdown Structure
Whole Person Concept	The guiding philosophy of the security clearance process.
Wizard War	Slang for the ongoing, cat-and-mouse game of technological one-upmanship in Electronic Warfare.

Book Recommendations and Further Reading:

This book was your map to the terrain. The resources on this list are the next step in your journey. They are the deep-dive intelligence reports, the legendary histories, the strategic analyses, and the ongoing briefings that will help you think and speak with the depth and context of a seasoned professional.

Reading is a form of reconnaissance. This list is not an exhaustive bibliography, but a curated professional library designed to build your mindset and deepen your understanding of the defense and aerospace world.

Foundational Histories & The Birth of the Modern Industry

These books explain the "DNA" of the industry; the historical events and legendary figures who forged its culture.

Skunk Works: A Personal Memoir of My Years at Lockheed by *Ben Rich & Leo Janos*. An insider's account of operating at the absolute cutting edge of aerospace technology on some of the world's most secret projects during the Cold War.

Failure Is Not an Option: Mission Control from Mercury to Apollo 13 and Beyond by *Gene Kranz*. A riveting first-hand account of what it means to build and operate a high-consequence, "flight-critical" engineering culture.

The Right Stuff by *Tom Wolfe*. A brilliant look at the culture, the risk, and the human element of the test pilot world that pushed the boundaries of early aerospace engineering.

Arsenal of Democracy: FDR, Detroit, and an Epic Quest to Arm an America at War by *A. J. Baime*. The dramatic narrative of the public-private partnership that created the modern defense industry.

The Wizard War: British Scientific Intelligence 1939-1945 by *R. V. Jones*. The original story of the "cat-and-mouse" game of Electronic Warfare; the battle of wits between radar, jammers, beams, and countermeasures, that continues to this day.

Command of the Air by *Giulio Douhet*. One of the foundational (and controversial) texts of air power theory, arguing for the supremacy of strategic bombing, a concept that directly led to the creation of strategic bombers like the B-21.

Blind Man's Bluff: The Untold Story of American Submarine Espionage by *Sherry Sontag and Christopher Drew*. For anyone interested in the Sea Domain, this is the real-life Hunt for Red October. It is a riveting, declassified history of the high-stakes intelligence operations conducted by U.S.

submarines during the Cold War. It provides incredible context for why acoustic stealth became the single most important design driver for submarines.

The Making of the Atomic Bomb *by Richard Rhodes.* This Pulitzer Prize-winning book is more than a history of a weapon; it is the single best account of the largest, most complex, and most secret engineering project ever attempted. It's a comprehensive case study of the intersection of physics, engineering, project management, and national security, and it details the birth of the "high consequence" mindset and the personnel security system itself at places like Los Alamos.

Understanding the Modern Strategic Landscape

These books explain the "why" behind the headlines and the technological megatrends shaping the future battlefield.

The Kill Chain: Defending America in the Future of High-Tech Warfare *by Christian Brose.* The single most important modern text for understanding the "Third Offset Strategy" and the military's shift toward autonomous, networked, and AI-enabled systems.

Ghost Fleet: A Novel of the Next World War *by P. W. Singer and August Cole.* A fictional, but deeply researched, "techno-thriller" that reads like a playbook for future conflict. It is a fantastic and accessible way to understand the vulnerabilities and capabilities of the modern, multi-domain force.

LikeWar: The Weaponization of Social Media *by P. W. Singer and Emerson T. Brooking.* A crucial read for understanding the "information operations" and Cyber Domain aspects of modern conflict, where the battlefield is increasingly digital.

The Pentagon's Brain: An Uncensored History of DARPA *by Annie Jacobsen.* A deep dive into the legendary and secretive agency responsible for countless technological revolutions, from the internet to GPS and stealth.

Army of None: Autonomous Weapons and the Future of War *by Paul Scharre.* A thoughtful and deeply informed analysis of the ethical and engineering challenges of artificial intelligence and lethal autonomous systems.

The Hundred-Year Marathon: China's Secret Strategy to Replace America as the Global Superpower *by Michael Pillsbury.* To be a great defense engineer, you must understand the strategic context that drives the requirements for the systems you build. This book is one of the most influential (and debated) modern texts on the long-term strategic competition that is the primary focus of the Pentagon today.

This Is How They Tell Me the World Ends: The Cyberweapons Arms Race *by Nicole Perlroth.* A terrifying and essential read for anyone interested in the Cyber Domain. This is the definitive, journalistic account of the global market for software vulnerabilities (known as "zero-days") and the dawn of state-sponsored cyberwarfare. It explains the "why" behind the immense focus on cybersecurity for weapon systems.

Thinking Like a Great Engineer

These are books about the craft itself; the timeless principles of design, problem-solving, and managing complexity.

To Engineer Is Human: The Role of Failure in Successful Design *by Henry Petroski.* A classic that teaches the critical mindset of professional paranoia and the importance of learning from past engineering failures.

Thinking in Systems: A Primer *by Donella H. Meadows.* A rich text offering insight into systems thinking, a crucial skill for any engineer working on complex systems of systems.

The Design of Everyday Things *by Don Norman.* While focused on commercial product design, this is a comprehensive examination of human-centered design and ergonomics; the art of making technology usable for human beings, a core challenge in designing soldier systems.

The Checklist Manifesto: How to Get Things Right *by Atul Gawande.* An incredibly powerful argument for the importance of process and checklists in managing high-stakes, complex tasks, from surgery to engineering.

Surely You're Joking, Mr. Feynman! *by Richard Feynman.* A lesson in the power of relentless curiosity, first-principles thinking, and the joy of solving a hard problem, from one of the 20th century's greatest scientific minds.

The Soul of a New Machine *by Tracy Kidder.* This Pulitzer Prize winner is the timeless story of a team of engineers at Data General in the late 1970s, racing to design a new minicomputer. It is, perhaps, the best book ever written on the culture of a high-pressure engineering project: the long hours, the inside jokes, the moments of frustrating failure and brilliant breakthrough. It perfectly captures the spirit of what it means to be on a design team.

Turn the Ship Around!: A True Story of Turning Followers into Leaders *by L. David Marquet.* This is a modern classic on leadership, told through the story of a U.S. Navy submarine commander. Marquet's "leader-leader" model, which pushes responsibility and decision-making down to the lowest possible level, is a powerful lesson for any aspiring lead engineer or manager. It is a guide to creating a culture of empowered, proactive problem-solvers.

Your Professional Daily Briefing (Online Resources)

A professional stays current through a daily habit of reading. These are the premier open-source resources to keep on your radar.

Daily & Weekly News:

Breaking Defense: Excellent for high-level analysis of budgets, strategy, and technology.

Defense News: The "paper of record" for the industry, covering contracts, policy, and global developments.

War on the Rocks: A premier outlet for deep, nuanced analysis of national security, military strategy, and technology.

The War Zone: A section of *The Drive*, known for its deep technical dives, analysis of satellite imagery, and "scoops" on new and exotic hardware.

Domain-Specific Resources:

Air & Space: Aviation Week & Space Technology (the gold standard), Air & Space Forces Magazine.

Sea: U.S. Naval Institute (USNI) News & Proceedings, The Maritime Executive.

Land: Army Times, National Defense Magazine.

Cyber: *C4ISRNET*, *WIRED*'s security section.

The Deeper Dives:

Center for Strategic and International Studies (CSIS): Their publications on space, AI, and military technology are superb.

The RAND Corporation: The original "think tank," produces incredibly deep and data-driven reports on every aspect of defense and technology.

The MITRE Corporation: A non-profit FFRDC that publishes excellent, technically deep explainers on complex defense-related topics.

About the Author

Alec Milner is a Research and Development Mechanical Design Engineer at Sandia National Laboratories and a systems integrator with deep experience in the aerospace and defense industry. His career has taken him through the full engineering lifecycle, ranging from the gritty, hands-on testing of tactical vehicles at government proving grounds to the abstract, high-consequence design of advanced strategic systems within the national lab complex.

Throughout his work on next-generation space and missile programs, he has collaborated with interdisciplinary teams to develop complex hardware, spearheaded multi-million-dollar component redesigns, and directed the formal qualification of systems intended for use in the most extreme environments imaginable.

This journey from the mud of the test track to the sterile clean rooms of national labs revealed a fundamental gap: the immense difference between academic experience and the unwritten, high-stakes cultural and professional rules needed to succeed in the world of national security.

He holds a Bachelor of Science in Mechanical Engineering with a concentration in Aerospace Engineering from Johns Hopkins University and is completing a Master of Science in Engineering from Purdue University focused on the intersection of technology and defense strategy.

Alec wrote this book after years of mentoring junior engineers and watching them struggle with the same intimidating "black box" he faced. He created it to be the map he wishes he'd had: a comprehensive, no-nonsense "insider's guide" designed to demystify the industry and empower the next generation of engineers to build careers of purpose, integrity, and impact.

www.ingramcontent.com/pod-product-compliance
Lightning Source LLC
Chambersburg PA
CBHW071537210326
41597CB00019B/3032

* 9 7 9 8 9 9 4 2 8 9 7 1 6